ALL·WHEEL·DRIVE
High·Performance
Handbook

Jay Lamm

Motorbooks International
Publishers & Wholesalers ®

to Michael Lamm

First published in 1990 by Motorbooks International Publishers & Wholesalers, P O Box 2, 729 Prospect Avenue, Osceola, WI 54020 USA

Motorbooks International books are also available at discounts in bulk quantity for industrial or sales-promotional use. For details write to Special Sales Manager at the Publisher's address

Library of Congress Cataloging-in-Publication Data
Lamm, Jay William.
    All-wheel-drive high-performance handbook / Jay William Lamm.
      p. cm.
    ISBN 0-87938-419-0
    1. Automobiles—Four-wheel drive.   I. Title.
TL235.6.L36 1990 90–5960
629.222—dc20 CIP

On the front cover: A new era of sophistication in all-wheel-drive performance was hailed by the debut of the 1989 Porsche Carrera 4 with its electronic control system for the fully automatic drive. *Porsche*
On the back cover: The 1989 Audi 90 Quattro GTO IMSA racer bringing all-wheel-drive performance to the world's racetracks—and winning in convincing style. Details of all-wheel-drive coupling. Mazda rally car in action.

Printed and bound in the United States of America

Contents

Acknowledgments

The following people and organizations generously donated time and effort to the completion of this book: Mike Aberlich; Ian Adcock; Fred Aikins; Mike Allen; Rich Allen; American Honda Motor Co. Inc.; Jay Amestoy; Kurt Antonius; Bobby Archer; Tommy Archer; Archer Brothers Racing; Audi AG; Audi of America; Austin-Rover Group; *Automobile* magazine; *Automobile Quarterly* magazine; *Automotive Engineering* magazine; Bill Baker; Richard Baron; Dieter Basche; Basic Living Products Inc.; Brian Berthold; BMW of North America; Dick Bonheim; Borg-Warner Automotive Inc.; Bosch AG; Jim Bowman; Klaus Brandt; Glenda Brown; Warren Brown; John Buffum; Buick Motor Co.; Ed Bumbalo; *Car and Driver* magazine; CBS Communications Inc.; Chevrolet Motor Co.; Chevrolet Public Relations; Chrysler Corp.; John Clagett; Kevin Clemens; Club Team Lotus; Competition Ltd; Mike Cook; *Corvette Fever* magazine; Jeff Cushing; Doug Davis; Steve Davis; Lucille DeLorenzo; Kim Derderian; Paul Devlin; Diamandis Communications Inc.; Jack Doo; Ben Dunn; Lisa Dunn; Geno Effler; Andrew Ferguson; Ferrari of North America; Fiat Auto USA Inc.; Ford of Europe; Ford International Public Affairs; Ford/Michelin Ice Driving School; Ford Motor Co.; Ford Overseas Division; *Four-Wheel and Off-Road* magazine; Fred Mackerodt Inc.; Freeman/McCue Public Relations; Elizabeth Gardiner; Geltzer & Co.; General Motors Truck and Coach Division; Michael Geylin; Goodyear Tire & Rubber Co.; Group 44; Rod Hall; Hank Forssberg Advertising; Hurley Haywood; Dave Hedrick; Gene Henderson; High Desert Racing Association; Mike Hohn; Josef Hoppen; *Hot Rod* magazine; Deke Houlgate; Ellen Houston; Indianapolis Motor Speedway Co.; Indy 500 Photos; International Motor Press Association; International Motor Sports Association; Terry Jackson; Jaguar Cars Ltd.; Jeep/Eagle Division; John Adams Associates; Mike Kaptuk; Kermish-Geylin Public Relations; Mike Knepper; Kristy Koppel; Dave Krupp; Charlie Lamm; JoAnne Lamm; Michael Lamm; Robert Lamm; Lamm-Morada Publishing; Ed Lechtzin; Leo Levine; Libra Racing; Guy Light; Light Performance Works; Jean Lindamood; Jackie Loeb; Dennis Lopez; Lotus Cars of America; Jean-Paul Luc; Joe Mack; Fred Mackerodt; Alex Maduros; Manning, Selvage & Lee; Dick Maxwell; Mazda Information Bureau; Mazda Motors of America, Inc.; Dan McCue; Eileen McDonald; Tom McDonald; Mary McElyea; Kenneth McKay; Rita McKay; Martha McKinley; Mercedes-Benz of North America; Otis Meyer; Michelin Tire Co.; Mike Michels; Rob Mitchell; Mitsubishi Motor Sales of America; Mitsubishi News Bureau; Mike Moran; *Motor Trend* magazine; Tetsuo "Ted" Nagase; Maria Nahigian; Maureen Nelson; Nissan Motors; Jan P. Norbye; Brett O'Brien; Bill Pauli; Karen Penn; Peugeot Motors of America Inc.; Pirelli Tire Co.; Maria Polleiner; Pontiac Motor Co.; Pontiac Public Relations; *Popular Mechanics* magazine; Porsche Cars of North America; Paul M. Preuss; Range Rover North America Inc.; *Road & Track* magazine; Graham Robson; Rod Hall Racing; Rod Millen Racing; Kari St. Antoine; Adam Saul; Alan Shaffer; Shelby Automobiles; Tom Smitham; Society of Automotive Engineers; David Solar; Southern California Off Road Enterprises; Sports Car Club of America; Diana Sprig; Ingmar Spruz; Bob Steele; Debbie Stern; Subaru of America Inc.; Vincent Sweeney; Tamarack Rallysport; Tartan Advertising; Team McPherson; Sonny Tippe; Toyota Motor Sales USA; Carol Traeger; Ed Triolo; U.S. Auto Club; Robby Unser; Vagabond Travel Inc.; C. Van Tune; Paul Van Valkenburgh; Jason Vines; Volkswagen of America; Washington Automotive Press Association; Larry Weis; Christy Whalen; Whole Earth Electronics Inc.; Hal Williams; Bill Wilson; Gary Witzenburg; David Yu; Paul Zazarine.

Chapter 1

All-Wheel-Drive Theory and Practice

The upper performance limits of any car, whether it's a Volkswagen Beetle or an all-wheel-drive rally car, are determined by traction. This applies not just to forward acceleration but to cornering and braking as well. Forward acceleration is simply the most common example.

Many cars can't crank out enough torque to break loose their driving wheels on a good, clean surface, so all the energy their engines make goes into forward acceleration. On slippery surfaces or with stronger engines, though, more torque may be available than the tires can deliver to the road.

The Porsche Carrera 4 was introduced in 1989 with sophisticated electronics to control the fully automatic all-wheel-drive system. Sesno Sensors provide data on wheel speeds and on the forces acting on the chassis in cornering, accelerating or braking. Two multi-disc clutches are engaged by the drive system computer as needed. The first controls torque division front to rear; the second acts as a limited-slip differential between the rear wheels. Clutch actuation comes from hydraulic pressure and fast-acting solenoid valves. *Porsche*

On the limits of even all-wheel-drive adhesion, Walter Rohrl's Audi Quattro Turbo Coupe charging up Pikes Peak. Audi set the stage for all-wheel-drive performance with the Quattro, winning in rally and hillclimb competition in the days when most racers scoffed at the need for power to all four wheels. *Audi of America*

Audi's second coming, the Quattro Trans Am winner. Most people were skeptical of all-wheel-drive until Audi proved its worth on the rallying circuits of the world. Then they said it couldn't be done in road racing. Audi's 1988 Trans Am season proved the value of all-wheel-drive on both wet and dry pavement. *Audi of America*

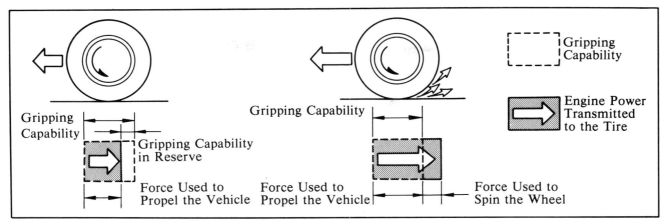

If the force that moves a car forward isn't greater than the tractive ability (gripping capability) of the car's tires, there's no problem. If the force is greater, wheel spin occurs and the car doesn't move. *Toyota*

When that happens, the drive wheels spin while the car itself hardly moves. As for braking and steering, once the tires lose traction and start skidding, the car has exceeded its ability to do what's being asked of it. In other words, no matter what modifications in hardware you make or what nifty maneuvers you try behind the wheel, there has to be enough grip in the tires to translate your efforts into useful motion.

To understand and use an all-wheel-drive automobile, you must actively consider that grip equals performance. You must take a serious look at grip, or traction, and ask what it is, how it's gained or lost, and how it can and can't be manipulated toward your performance goals.

Traction

For our purposes, traction is used to describe how well a tire grips the road. That grip is used to move the car forward under acceleration, to move it to the side during cornering or to slow it down during braking. The more traction available, the faster any of these functions can be performed.

With lots of traction available, a fast car can accelerate with about the same amount of force as the force of gravity (g) that's pulling it to the ground. Therefore, accelerating at a force equal to that of gravity gives a figure of 1 g, accelerating with half the force of gravity would be measured as 0.5 g and so on.

If less traction is available—say there's oil on the road—the engine still has enough power to move the car forward at 1 g, but the tires no longer have enough grip. Let's say the tires can only provide enough grip for an acceleration of 0.5 g. The extra torque above and beyond that is wasted on wheel spin instead of forward motion. Furthermore, once the tires start spinning and contact with the road is essentially lost, there's almost no forward motion to the car.

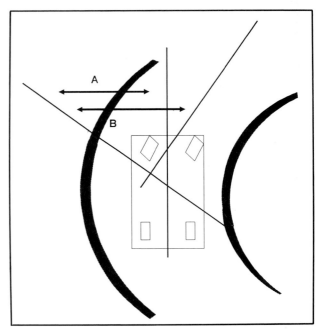

Drift is the angle between the actual direction of travel at a given moment and the direction that the car is pointing (A). Front drift equals the total drift plus or minus the angle of the front wheels (B). *Tom Smitham Graphics*

So far this is all simple stuff, but the relationship between tire and road surface is more complicated. All driving maneuvers involve an element of drift (some slippage between the road and tire). In cornering, for example, a drift angle reflects the difference between the direction the tires are pointing and the direction the car is traveling at that instant. The rear tires and the vehicle have the same drift angle; the front tires' drift angle is measured by adding or subtracting their own angle relative to the vehicle's drift angle.

A tire's chemical make-up, or compound, determines its grip in part. Friction-heated racing tires can become sticky enough to pick up cigarette butts and rocks.

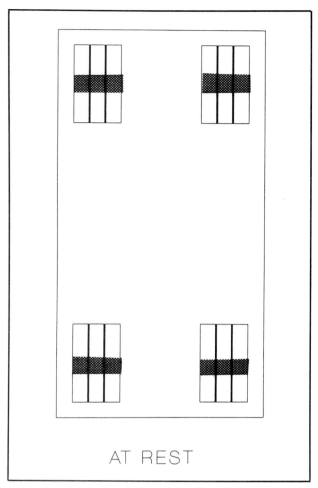

AT REST

Assuming perfect weight distribution, a car at rest sits on four equal contact patches. Under braking the front contact patches are considerably larger than the rear's, however, while the opposite is true under acceleration. *Tom Smitham Graphics*

Moving tires aren't hooked up as we'd like to believe. There's even a good deal of slippage between rubber and road when the car is traveling at a steady speed down the highway. We don't need to worry about this slippage for most of our discussion, but keep it in mind.

Overall traction is determined by the surface of the road, the stickiness of the tires, the weight carried by the tire, and the size of the contact area between tire and roadway. All of these variables affect the amount of traction available at any given moment. The first two are the easiest to define, so let's deal with them to start with.

The grip offered by a road surface is most obviously determined by whether or not a foreign material lies between the tire and the pavement. Snow, grease, water or almost anything else that comes between the tread surface and the road tends to reduce the available traction—by just how much is discussed in Chapter 6.

A tire's stickiness is determined by its chemical make-up, or compound: the softer the rubber, the more traction it offers. A tire provides traction because, on a microscopic level, its rubber drapes into openings in the road surface and pushes against them. The softer the tire, the more it presses down into the minute valleys of the road and the more area it has to push against.

Tire temperature also affects adhesion: hot tires are softer, and therefore stickier, than cold ones. Of course, soft tires wear out more quickly

than hard ones, so a compromise must always be made between optimal grip and longevity.

The second two variables in the traction equation—the weight being carried by the tire and the size of the contact area—are inseparable. The term contact patch or tire patch refers to the interacting area between tire and road. The more weight you put over a tire, the larger its contact patch for a given inflation and tire size. The size of the contact patch determines available traction simply because it determines how much frictional surface is being employed between the tire and the road.

At rest, a car with fifty-fifty weight distribution and four equal-sized tires sits on four equal-sized contact patches. Each contact patch of a regular automobile is only about the size of a tennis shoe sole, so making the most of it is imperative.

Under acceleration, inertia transfers weight to the rear of the car and the rear contact patches

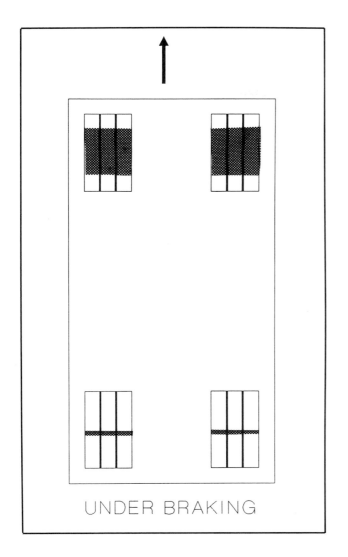

UNDER BRAKING

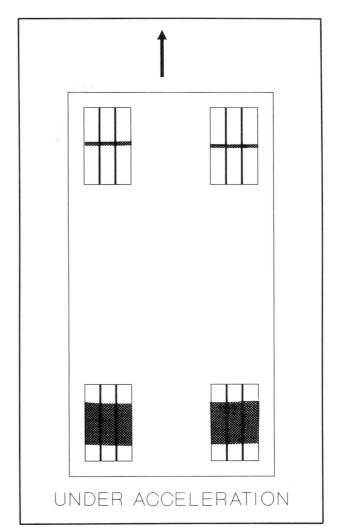

UNDER ACCELERATION

Front-wheel-drive performance cars like the Pontiac Grand Prix are hampered by weight transfer under acceleration. The six-cylinder Grand Prix will easily spin its tires on takeoff. *Pontiac*

Fortunately, these Ford-Lotus Cortinas have rear-wheel-drive. Weight transfer has lifted the inside front wheel and rolled the body toward the outside of the corner. *Ford of Europe*

become larger than the front contact patches. At this point, the rear tires can provide more traction than the front ones can. This is all well and good with rear-wheel-drive, because the back tires push the car forward.

Some people ask why a front-engine, front-wheel-drive car can out-accelerate a rear-drive car on snow. Isn't getting traction even more of a problem on snow than on dry pavement? The trick here is that overall acceleration rates on snow are much lower, so weight transfer is a much smaller variable of the equation. The added weight of the engine over the car's drive wheels more than compensates for the grip lost to weight transfer.

But the question about front-wheel-drive cars on snow brings up an important point. On dry pavement, weight transfer under acceleration is a problem for front-drivers. If enough weight is unloaded off the front tires, the front contact patches can become so small that they're no longer able to get the engine's full torque to the ground. All else being equal, a front-drive car will suffer wheel spin appreciably earlier than will a rear-drive car. For the best of both worlds, then, you have to turn

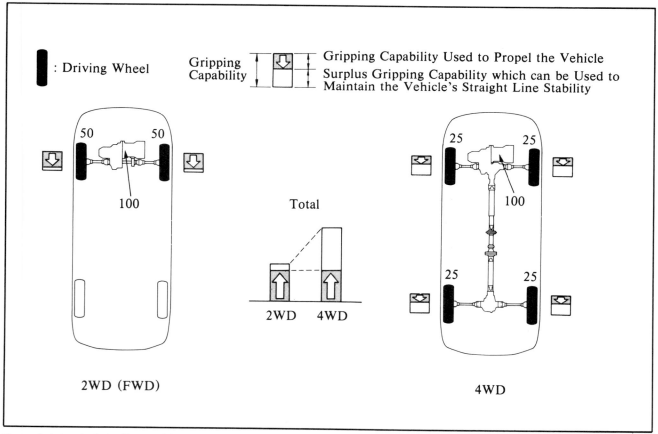

To maintain straight-line stability, a car needs surplus grip at all four tires. With two-wheel-drive, there's less surplus on the driving wheels than with all-wheel-drive. *Toyota*

to front-*and*-rear-wheel-drive—but we're getting ahead of ourselves.

Let's move on to braking. Under braking, weight shifts to the front and the front contact patches become larger. Since the front tires are now being asked to haul more weight down from speed than are the rear tires, the added traction available from them is a good thing. And since the front tires are also being asked to steer if the car is braking for a corner, their larger contact patches give added grip right where it's needed.

If the driver of a rear-wheel-drive car were to suddenly get on the gas again at the entry to the corner, he or she would have to do it gently enough to shift weight to the rear before giving the car full throttle. Otherwise the demand on the unloaded rear wheels would exceed the wheels' tractive ability, and the rear of the car would slide to the outside of the corner. (This can be helpful in setting the car to the proper attitude for a corner, a driving maneuver discussed later on.)

Weight transfer doesn't work only in the fore-and-aft direction, but from side to side as well. On a fast corner the inside rear wheel will have the smallest contact patch—it may even be entirely off the ground—as the car brakes into the turn. As the driver powers out of the turn, the inside front wheel has the smallest contact patch—and it, too, may leave the ground. Needless to say, a tire that's

off the ground isn't contributing anything to the handling or stability of the car.

By maximizing the available traction through weight transfer, a skilled driver can tailor the traction of individual tires to his or her own needs. More traction can be put up front for cornering, more on the rear for acceleration and so on. All high-performance driving is really just a study in maximizing traction; correctly apportioning it where and when it is needed.

What dangers are to be dealt with when juggling traction from tire to tire? An unloaded front end means that steering and braking will not be at their best, and an unloaded rear end means that the tail end won't be doing all it can to provide lateral stability.

At the heavily loaded end, meanwhile, is the very real danger of saturating a tire. A tire can do only a given amount of work. It can generate only so much grip. If the maximum tractive ability of the tire is already being used to stop, for example, no grip is left to help steer. Similarly, a tire that's using all its tractive ability for forward acceleration can't provide lateral stability at the same time. Overloading a tire is just as dangerous as underloading a tire if you don't know how to deal with it.

Saturation of the front tires, often caused by too much simultaneous steering and braking, leads to understeer: the front tires slide and the car goes straight when it's supposed to turn. Saturation of

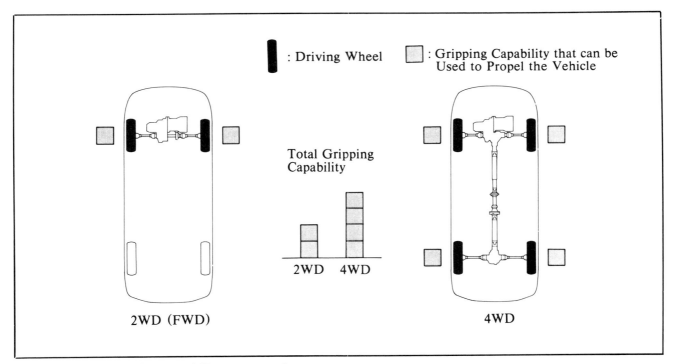

: Driving Wheel   : Gripping Capability that can be Used to Propel the Vehicle

Total Gripping Capability

2WD   4WD

2WD (FWD)   4WD

Twice the total gripping capability means twice as much force can be used to propel the vehicle forward.
Toyota

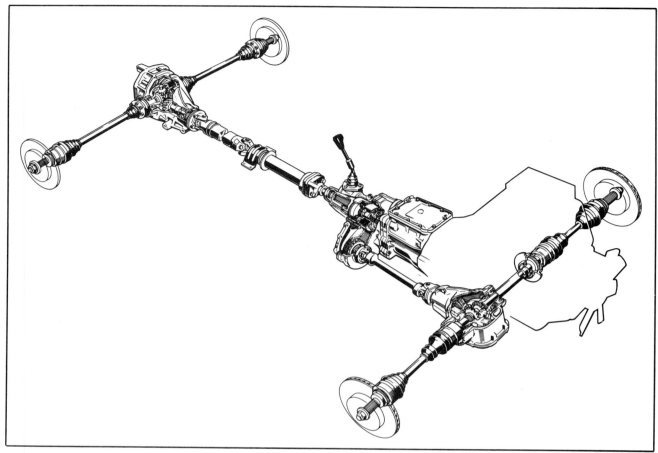

Transfer case, center differential, front driveshaft, front differential and front axle shafts were added to convert the Ford Sierra to all-wheel-drive. All add weight, cost and complexity. *Ford of Europe*

the rear tires leads to wheel spin under acceleration and oversteer: the tail comes out during fast cornering.

Advantages of all-wheel-drive

Now let's take a look at how all-wheel-drive makes the most of a car's available traction. First, most people know that all-wheel-drive offers better forward acceleration on slick pavement. But why is that so?

We've been using grip and traction as more or less undefined quantities until now. A precise measure of traction is called the coefficient of friction, usually represented by the symbol k.

Say you have a 100 lb. block of cement sitting on a smooth floor. If it takes 50 lb. of force to get the block to move, the block's static coefficient of friction is figured this way: 50 lb. (the force required to move the block) divided by 100 lb. (the force of gravity holding the block to the floor) equals 0.5 k (the static coefficient of friction).

The force required to *keep* the block moving will be less. Let's say it takes just 10 lb. of force to keep the block going once it's started. In this case, the *sliding coefficient of friction* is 10 lb. divided by 100 lb., or 0.1 k.

Recall that tires are not rigidly hooked up with the road. In fact, when cornering quickly, a car's rear drift angle can easily be 15 degrees or more before the driver perceives any rear-end slide at all. When discussing the coefficient of friction in relation to moving cars, then, we're talking about a point somewhere between the sliding and static coefficients. Some people prefer to call this figure the coefficient of adhesion, but this book will refer to it simply as the coefficient of friction. Whatever you call it, this figure will be much closer to the static coefficient, which is always higher than the sliding coefficient. (This is why once a wheel begins to slide it offers considerably less traction than it did when it was hooked up.)

Now let's apply this theory to a car. Our test car will weigh 2,000 lb. and will have perfectly equal weight distribution front to rear and side to side (conveniently enough).

Let's say this car's trying to move across rough ice. The coefficient of friction of a tire on rough ice is about 0.2 k, and it doesn't change much with speed. On most other surfaces, the coefficient of friction drops off as speed increases.

If we try to move our 2,000 lb. car with the tractive force of just one wheel, the best forward acceleration we can hope for is 0.05 g. Why? The trick here is that even though the car weighs 2,000 lb., only 500 lb. of that is over the driving wheel. The 500 lb. weight on the wheel multiplied by 0.2 k (the coefficient of friction) equals 100 lb. of force. This 100 lb. (the maximum force that can be used to move the car) divided by 2,000 lb. (the weight of the car) equals 0.05 g of acceleration.

No doubt you can see where this is all heading. If we make the car two-wheel-drive, the best we can hope for is 0.1 g, since 1,000 lb. are over the driving wheels ($1,000 \times 0.2 = 200$, and $200 \div 2,000 = 0.10$). Finally, if we plug in the numbers with all four wheels providing forward thrust, the best theoretical forward acceleration becomes 0.2 g, or twice as much as we had with two-wheel-drive, four times as much as with one-wheel-drive.

This is an imperfect model, but it's a sound demonstration of an underlying principle of all-wheel-drive: With all-wheel-drive, all the weight of the car, rather than just the weight over the wheels on the driven axle, can be used toward generating traction for forward acceleration.

So what's wrong with this model? Plenty. For one thing, it doesn't take weight transfer into account, though on ice the acceleration levels are so low that weight transfer is almost negligible. For another, it assumes that the all-wheel-drive car has its driving torque split fifty-fifty between the front and rear axles, which many do not. It also doesn't account for any differential action—the unequal torque distribution that comes from having two or three differentials, as most all-wheel-drive street vehicles do. It further assumes that the driver is skilled

The all-wheel-drive Mitsubishi Galant GSX can perform maneuvers on dirt that would have front-wheel-drive and rear-wheel-drive cars sliding to the outside of the turn. *Mitsubishi of America*

enough to accelerate just at the limit of traction. It doesn't take into account that some tractive ability must be left in the tires to provide lateral stability, and so on and so forth. But the basic principle remains sound regardless.

This model shows that all-wheel-drive cars have a big advantage in forward acceleration, but all automobiles use all four wheels to stop and corner. What's the advantage of all-wheel-drive in these areas?

One advantage comes from avoiding saturation, the condition wherein one tire is asked to do

Mitsubishi's Eclipse GSX comes loaded to the gunwales with a double-overhead-cam 16 valve turbocharged engine, all-wheel-drive, 16 inch wheels, power mirrors, a six-way seat, an oil cooler and more. Its base price reflects all the mandatory goodies. *Mitsubishi of America*

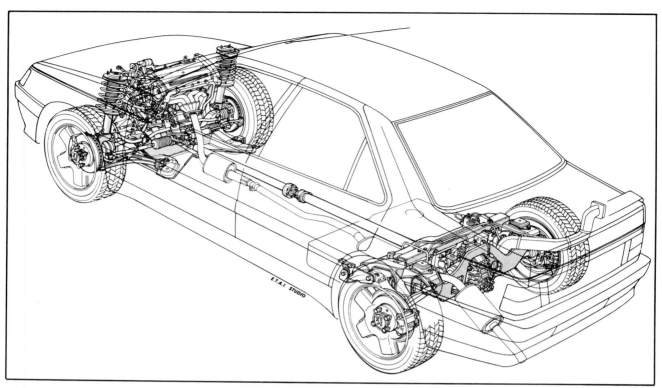

The entire driveshaft, rear axle and rear suspension systems of the Peugeot 405 16X4 had to be designed at extra cost. They also add weight and present more components for gremlins to invade—that's the price of all-wheel-drive. *Peugeot of America*

more than it's capable of. In our earlier example of saturation, the rear wheels of a rear-wheel-drive car were asked to provide forward thrust (acceleration) and side thrust (lateral acceleration) at the same time. If the sum of the two demands exceeds the tractive ability of the driving wheels, traction will be lost and the rear tires will skid.

If, however, we reduce the amount of forward thrust being delivered by the rear tires by also using the front tires for that purpose, the tractive ability freed up at the rear can be put toward lateral acceleration. The sum of the forward thrust and side thrust might now be less than the maximum tractive ability of the tires, and the skid is avoided.

We usually think of lateral acceleration in terms of cornering force, but it's also used simply to keep a vehicle traveling in a straight line. Therefore, an all-wheel-drive vehicle is also more likely to remain traveling straight under hard acceleration than is a two-wheel-drive vehicle.

Lateral acceleration is not the only place an all-wheel-drive vehicle's added stability comes from. When one tire of a two-wheel-drive vehicle loses traction, the driving force is momentarily concentrated on the other side of the axle. The result is sideways motion, or yaw, which at high speeds can be serious enough to produce a slide. In a front-wheel-drive car, the nose of the vehicle will rotate

toward the side of lost traction, while a rear-wheel-drive car rotates away from the side of lost traction. With all-wheel-drive, the yawing action induced by both front and rear drive wheels effectively cancels itself out, the result being continued motion in a straight line.

An all-wheel-drive vehicle with a locked or limited-slip center differential will also offer better stability in emergency or fast-driving situations that lead to lost traction at one end of the car. With a two-wheel-drive vehicle, when traction is lost at the drive axle the only major force applied to the vehicle is inertia. Inertia pushes the car in the direction it was originally traveling. With all-wheel-drive, the second drive axle can continue to add force in the direction the vehicle should be traveling, which is often not the direction it was previously traveling.

Disadvantages of all-wheel-drive

All these benefits make all-wheel-drive look like a technology without drawbacks, but it does have disadvantages. Basically there are three: cost, complexity and weight.

The cost of an all-wheel-drive system varies greatly from application to application. Aside from the dollar value of the hardware itself, there are added tooling expenses, added research and development costs, and, in some cases, patent licenses that have to be paid for. All these expenses are passed on to the consumer and are reflected in the car's sticker price. The pass-along costs depend greatly on the sophistication of the driveline and the projected build volume of the vehicle. Because the levels of available research and tooling are constantly rising, the added cost of all-wheel-drive has fallen since the early eighties and will continue to fall for some time to come.

Another item that inevitably adds to the sticker price is the simple exotic appeal of all-wheel-drive. Many buyers interested in all-wheel-drive vehicles have their hearts set on the system beforehand, and dealers feel that they're willing to pay whatever it takes—within reason—to get it. It also doesn't help that these buyers are generally auto enthusiasts who are willing to pay more for a car than Joe Average would pay.

The cost of all-wheel-drive increases still more because few systems are offered on otherwise stripped-down models. All-wheel-drive is usually available only with top-of-the-line performance, handling and trim packages.

The added weight of all-wheel-drive, commonly between 100 and 200 lb., is another consideration against the technology. To some degree, all-wheel-drive means more than just added weight—it means added unsprung weight. Sprung weight is the mass of the car that's supported by springs, struts, shocks and so on. Unsprung weight is the mass that rides up and down with the wheels as they go over the pavement. The higher the ratio of unsprung to sprung weight, the more the vehicle's body will react to road imperfections. The driveshafts, constant velocity joints and so forth of an all-wheel-drive system all count as unsprung weight.

A final problem is that all-wheel-drive adds more components to the vehicle, increasing its complexity. With more components, more things can go wrong, although all-wheel-drive pieces are generally reliable. Added complexity also means higher friction losses at seals and constant velocity joints, plus a higher level of needed driver awareness if one or more of the components is under manual control, as with manually locking differentials.

Now that you're aware of the downside of all-wheel-drive, you should also know that aside from some initial sticker shock most drivers never notice these drawbacks. A hundred extra pounds is virtually unnoticeable on a street car. The reliability of all-wheel-drive components is usually excellent. The friction losses of the system are minor, at most, in practice. And the added unsprung weight is easily tuned out by suspension engineers.

If you can pay the price, then, nothing should discourage you from lining up all-wheel-drive for your next car.

Chapter 2

Making it Work

It would seem that making an all-wheel-drive automobile shouldn't be too hard. After all, all you'd really have to do is tie all four wheels together and then run power to the system at any point, right?

Wrong. The trouble with designing an all-wheel-drive system that way is that no two wheels of an automobile turn at exactly the same speed as the car goes down the road. Even going in a straight line, road irregularities cause one wheel to travel at a slightly different rate than the others. That's the minor problem. The major problem comes in turns.

When a car turns, each wheel covers a path of measurably different length. While the center of the car itself might follow the ideal line of the curve, the outside wheels must travel a longer distance than that line and the inside wheels a shorter one.

Meanwhile, both front tires prescribe larger arcs than do their mates at the back.

For example, let's take a vehicle with a rear track 4 ft. wide and move it through a 90 degree turn that's 100 ft. long when measured down the center of the road. The outside rear tire has to travel 102 ft., while the inside rear tire does the same corner in 98 ft. Connected together by a solid shaft, the inside tire would be forced to speed up to match the outside tire's faster revolution rate, or the outside one would have to slow down. Either way, the result would be a lot of noise and tire scrubbing as one wheel tried to keep up with the other. If the road surface were slippery, as with soft dirt or mud, the tires could slip freely enough that the driver might not notice all the noise. He or she would, however, notice that the vehicle was dis-

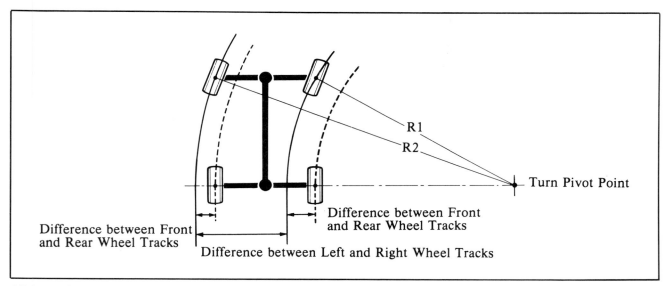

All four wheels travel different paths through a right turn. The right wheels cover less ground than the left wheels; the rear wheels cover less ground than the fronts. *Toyota*

tinctly reluctant to turn. If the surface were pavement, not only would the noise and tire wear be horrible, the axle wind-up caused by the force of the two wheels fighting each other would eventually break hardware someplace in the driveline.

Enter the differential.

Open differential

The traditional open differential is a curious-looking device that some demented genius figured out way back in the early nineteenth century. The basic idea is that instead of both drive wheels being connected by a single solid shaft, each wheel gets its own halfshaft with a bevel gear on the end opposite the wheel. This gear is held fast inside a hollow case, the cage. The axles' bevel gears are located on the flat sides of the cage. Attached inside the round walls are one or more pinion gears that mesh with the bevel gears. These pinion gears are free to rotate if they wish.

Finally, around the outside of the cage is a large ring gear driven by a pinion gear powered off the driveshaft. When the driveshaft turns, the ring gear rotates the cage in the same direction the wheels should be turning. If both wheels offer an equal amount of resistance, the pinion gears inside the cage mesh with the bevel gears on the axle shafts and drive each with equal force—in other words, the pinion gears don't rotate about their own axes. If one axle shaft offers more resistance than the other, though, the pinion gears turn about their own axes, rolling on the reluctant bevel gear and pushing the opposite wheel's bevel gear at a faster rate.

You can probably see the problem with this system. While it does a great job of allowing one wheel to rotate faster than the other around corners, it also causes the wheel with the least resistance to rotate the fastest. If one wheel is sitting on ice and the other on pavement, the wheel on solid ground—the one that should be turning to get the car moving forward—will offer more resistance, and the wheel on ice will rotate. The car won't go anywhere.

Taking off with one wheel on a slick surface isn't the only time an open differential causes problems. In hard cornering, for example, the inside driving wheel can become unloaded enough to lose traction on the pavement, and the driver will no longer have throttle control of the vehicle. When the tire loads up again, all sorts of nasty things can happen unless the driver is on top of the situation and ready to deal with the new-found grip. Similarly, if one drive wheel travels over a slick spot while the car is traveling straight, the momentary juggling of driving force from side to side can cause directional instability.

The first step toward controlling the unpleasant aspects of this otherwise helpful piece of hardware was the invention of limited-slip differentials.

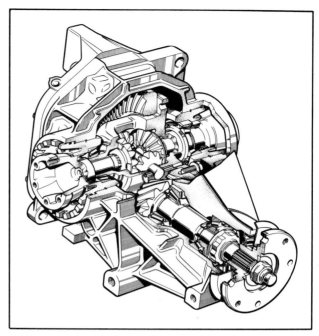

At the heart of the Ford Sierra's rear end is a simple open differential. Pinion gears in the middle don't rotate about their own axes until one bevel gear offers more resistance than the other. *Ford of Europe*

Limited-slip differentials

A number of limited-slip devices can be combined with the traditional open differential. The most common use some form of friction material tying together the differential cage and the reluctant axle shaft. The problem with these devices is that the friction material, just like a clutch in a regular driveline, can burn up with use over time. Other limited-slip differentials use mechanical principles—ratchets, ramps, balls and so on—to achieve the same goals. All of these limited-slip differentials are modifications of the original open differential, designed to provide lockup at a predetermined torque level or difference in axle shaft speeds.

Two completely different differentials—the viscous coupling and the Gleason Torsen—have come along since 1970, however, and both offer superior performance to traditional limited slips through clever mechanics. Because these don't use tacked-on gadgetry to provide limited-slip action, many people don't use the term limited slip in connection with them. That's the action they provide, however, and that's the term we'll use here.

These special limited-slip differentials are more expensive than regular limited slips, at least for the time being. That's partly because they can only be built under license, partly because the machinery for producing traditional limited slips already exists and is paid for, and partly because the characteristics of each unit must be specifically tailored to that unit's application.

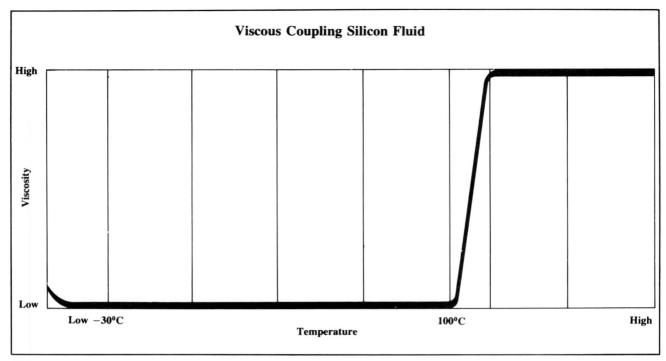

Viscous Coupling Silicon Fluid

Viscosity — High / Low

Temperature — Low −30°C / 100°C / High

Silicone oil remains at low viscosity until reaching a given temperature, then viscosity increases dramatically. The heat is provided by the friction of shearing oil molecules. *Volkswagen of America*

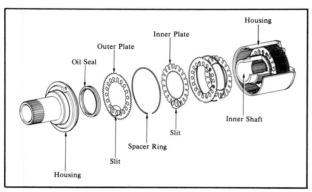
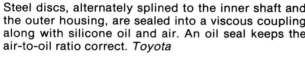

Steel discs, alternately splined to the inner shaft and the outer housing, are sealed into a viscous coupling along with silicone oil and air. An oil seal keeps the air-to-oil ratio correct. *Toyota*

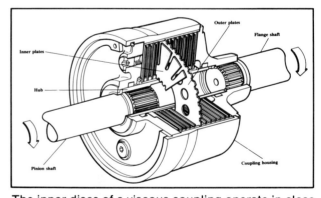

The inner discs of a viscous coupling operate in close proximity. Perforations, slots and holes provide added turbulence and shear force to silicone oil. *Volkswagen of America*

Viscous coupling

Harry Ferguson Research, the British company that pioneered all-wheel-drive for automotive applications in the fifties and sixties, was already the world leader in all-wheel-drive technology when it hit on a new differential concept in 1969. The Ferguson Formula equipment it was trying (with little success) to sell to manufacturers consisted of lots of gears, flywheels and one-way clutches. It worked fine, but the company was always looking for a simpler way to achieve the same goals.

The firm made a major step forward when a young visitor to its offices brought along a toy called Potty Putty. This was basically a wad of silicone oil that was pliable and liquid at rest, but stiffened when put under stress. It would sit like a soft lump on a desk, but bounced like a ball when dropped on the floor.

Ferguson's engineers immediately saw the potential of this oil. If it could be employed between two axle shafts, it would be inactive when the shafts were rotating at the same speed and therefore causing no stress—shear force—in the fluid.

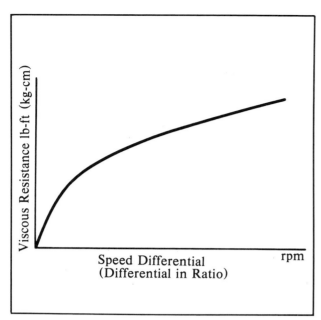

As the speed differential between discs in the coupling rises, horizontal, so do shear force and heat. Heat leads to increased oil viscosity and torque transmission, vertical. *Toyota*

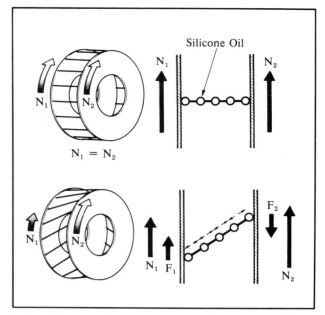

When disc speeds are equal, top, silicone oil exerts little force on discs. When discs rotate at different rates, bottom, viscous oil "drags" reluctant discs up to speed. *Toyota*

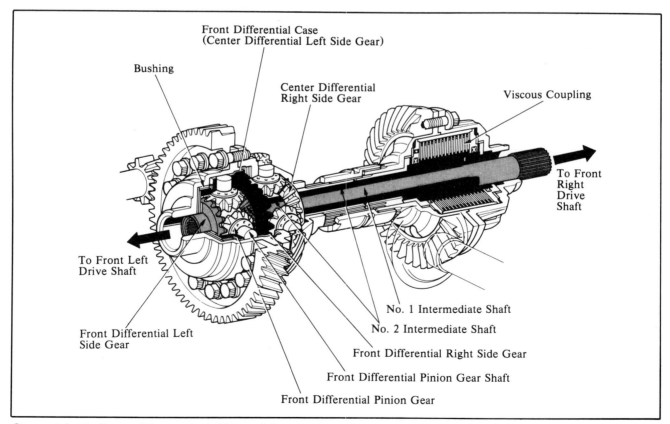

Concentric shafts provide compact differentials using viscous control. By running hollow shafts for the outer viscous coupling housing and inner viscous coupling shaft (No. 2 intermediate shaft), one even has room for a solid right driveshaft (No. 1 intermediate shaft) in the middle. *Toyota*

Should one shaft move faster than the other, though, the resulting shear force would immediately increase the oil's viscosity and make it act as a lock between the axles. Unlike with regular oil, heat causes the viscosity of silicone oil to increase dramatically. The shear forces create friction, which generates heat, which increases the oil's viscosity.

The way Ferguson employed the silicone oil was simplicity itself. The viscous coupling it evolved was a sealed can filled with the stuff, through the center of which ran a free-turning shaft. Alternately splined to the inside of the can and the outside of the shaft were a number of slotted steel discs. When the can and the shaft rotated together, the alternating discs ran at the same speed, and the oil remained fairly low in viscosity. But if the shaft and can rotated at different rates, the shear forces created between discs caused a frictional lock: the fluid stiffened and prevented the two components from turning freely in relation to each other.

By altering the number and design of the plates, the standard viscosity and amount of fluid, and the overall size of the unit, Ferguson engineers found they could adapt their viscous coupling to provide a virtually limitless range of torque transmission characteristics. The viscous coupling could be tailored to allow as little or as much overrun from one shaft to the other as desired and to lock up quickly or slowly. The engineers soon started including this new device in all their all-wheel-drive projects.

The viscous coupling has another advantage besides simplicity and effectiveness. Because there's usually no mechanical contact inside the viscous coupling, unlike friction-adding limited-slip devices it makes for less wear and tear on the driveline as a whole. For the same reason, the viscous coupling is virtually unbreakable. As long as it stays sealed and nothing goes wrong with the internal discs—which isn't very likely—the unit will continue to function.

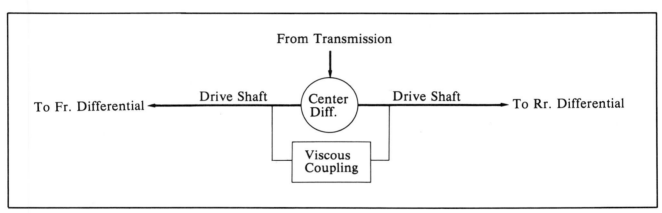

In an idealized center viscous coupling, power is taken off front and rear driveshafts on either side of an open center differential. When a speed difference "locks" the viscous coupling, the front and rear shafts are tied together. *Toyota*

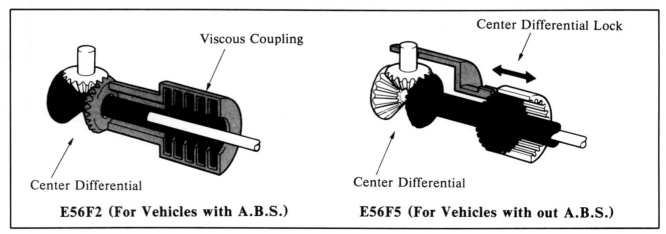

E56F2 (For Vehicles with A.B.S.)   **E56F5 (For Vehicles with out A.B.S.)**

A viscous center differential allows ABS to remain operational at all times. Some all-wheel-drive Toyotas get manual differential lock without ABS, viscous coupling with ABS. *Toyota*

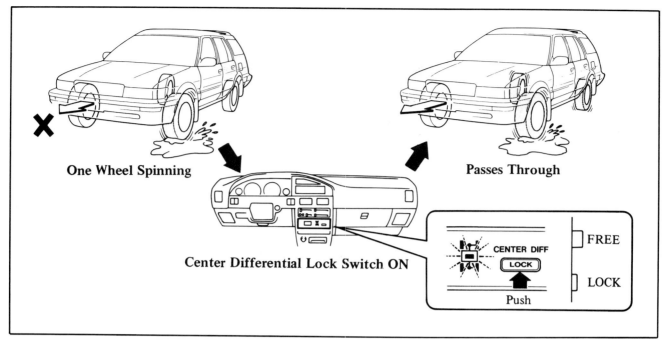

One Wheel Spinning

Passes Through

Center Differential Lock Switch ON

CENTER DIFF LOCK — Push — FREE / LOCK

Without limited slip, the center differential allows one slipping wheel to spin while others are bogged down.

Engaging the manual differential lock on the Corolla wagon feeds power to both axles. *Toyota*

Although usually there's no mechanical contact inside the viscous coupling, most units are designed so the internal plates come into physical contact under extreme conditions, transmitting torque even more efficiently than the fluid alone can. This condition is called humping and is another characteristic that can be built into the viscous coupling.

While the viscous coupling itself is a simple device, the installation of a viscous coupling into a driveline is a bit complicated. The most common layout is to place the viscous coupling unit next to a standard differential, its inner shaft lined up with the axle shafts or driveshafts. One axle shaft will be a hollow tube, and it also forms the outer case of the viscous coupling unit. The other shaft, rather than ending at its bevel gear, passes on through the center of the differential into the hollow tube of the opposite shaft. It ends by forming the center shaft of the viscous coupling unit. These concentric shaft arrangements are understandably tricky, and relatively expensive, to produce.

It's also possible to run geartrains off the axle shafts on either side of a standard differential and feed their power to the viscous coupling. This method takes up more space and involves more parts than the concentric shaft arrangement, however.

The first all-wheel-drive production car to use a viscous coupling was the AMC Eagle. With the later popularity of all-wheel-drive the technology has been taken up by a number of other manufac-

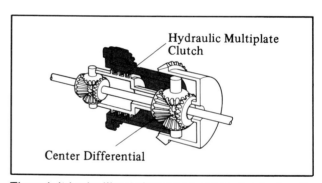

Hydraulic Multiplate Clutch

Center Differential

Though it looks like a viscous coupling, the hydraulic multiplate clutch unit in the Toyota driveline operates by pressing alternating discs of friction material together. *Toyota*

turers, such as Toyota, Mitsubishi, Volkswagen, Ford of Europe and others.

Gleason Torsen

The basic principle of the Gleason Torsen (for *Tor*que *sen*sing) is that of the worm and wheel. The most familiar example of the worm-and-wheel gear arrangement is a nut and bolt. The bolt is a worm gear, the nut a sort of worm wheel (its threads are on the inside instead of the outside, but don't let that throw you). Turning the worm gear (bolt) drives and worm wheel (nut) up the threads. Pushing the worm wheel doesn't move the worm gear at all, however. Visualize the pitch of the threads and you can see why. So far so good.

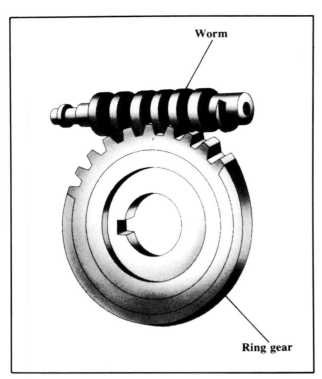

With the classic worm gear and worm wheel (ring gear), the gear can turn the wheel but not vice versa. *Audi of America*

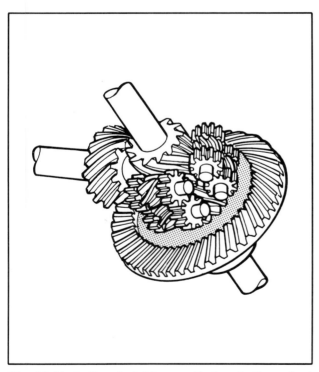

The trick's in the straight gears that mate the paired worm gears. The axle shaft's worm wheels would normally rotate the worm gears, but straight gears prevent it. *Peugeot of America*

Limited slip is provided by a Gleason Torsen rear differential in the Audi V8 sedan. Rights to the Torsen design are owned by Gleason Power Systems Division, Gleason Corporation. *Audi of America*

Now take the case of a push-type screwdriver. The threads are pitched more like the stripe on a candy cane than the threads of a bolt. Thus, pushing on the head of the screwdriver (worm wheel) does indeed turn the screwdriver shaft (worm gear).

Vernon Gleasman realized that by picking just the right pitch threads for worm gears and worm wheels—a pitch that allowed the worm wheels to impart just the correct amount of movement to the worm gears—he could make a limited-slip differential. Here's how it works.

The cage of the Torsen differential is about the same size and shape as that of an open differential. Inside, however, instead of pinion gears there are three sets of worm gears, and replacing the bevel gears at the ends of the axle shafts are worm wheels. Each set of worm gears has one gear paired to one shaft's worm wheel, and each pair of worm gears is connected by two straight gears.

When you turn the cage, what happens? If the worm gears had threads pitched like a screw's, they'd simply push on the worm wheels and the axles would go around. But remember, the idea of a differential is to allow a little difference in speed between the axle shafts. So the worm gears are pitched more like a candy cane's stripe than a screw's threads, and the worm wheel can turn the worm gears a little. This allows each axle some freedom relative to the differential cage.

Without being geared together, the worm gears would constantly rotate and lots of engine torque would be wasted. But the gears are geared together, and since both want to rotate in the same direction, their attaching straight gears prevent them from rotating at all. As before, all the movement of the cage goes into rotating the worm wheels on the axle shafts.

Now here's the tricky part. Because it has just the right pitch threads, the Torsen differential has

24

enough play to allow all the meshing gears some leeway. The mechanical action isn't perfect. It's just close enough that only a limited amount of axle shaft freedom is allowed—the amount needed to get through a corner smoothly.

Torsen differentials are an elegant solution, and they work like a charm. Formula One and CART racers use Torsens in their rear axles, which says something about their effectiveness and reliability.

Audi already offers Torsens in various applications, and experts predict they'll become more common in a big hurry. The Peugeot 405 Mi 16 x4, scheduled to arrive in America in 1991, already offered them as an option in Europe in 1989. Your next all-wheel-drive may very well come so equipped.

Unlike viscous couplings, Torsen differentials are straightforward replacements for a standard open differential. They don't need special concentric shafts or added geartrains to make them work. It's relatively common, in fact, for owners to retrofit their own automobiles with Torsen differentials.

Making all-wheel-drive work

Just as inside and outside wheels of a vehicle turn at different rates through a corner, the front and rear axles turn at different rates under all driving conditions. Even the marginal difference that usually exists between front and rear tire sizes is enough to create strain on the driveline if some sort of differential isn't used between drive axles. All but the most basic all-wheel-drive vehicles,

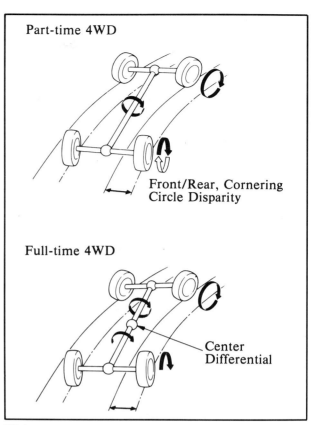

Part-time 4WD

Front/Rear, Cornering Circle Disparity

Full-time 4WD

Center Differential

Without a center differential, above, tires must slip to make up for rotational difference from front to rear. With a center differential, below, tires rotate only as much as necessary. *Toyota*

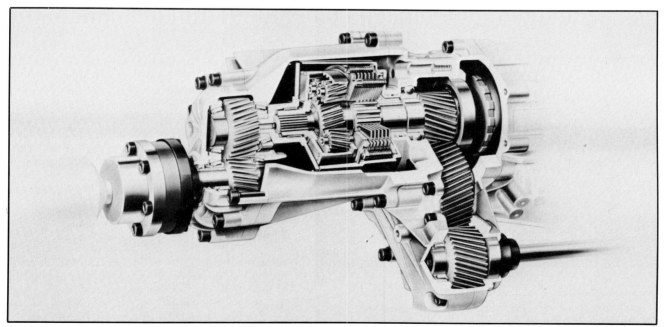

Although Torsen center differentials appear on some other Audis, the V8 sedan uses an electronically controlled multiplate clutch to provide limited slip in slick conditions. *Audi of America*

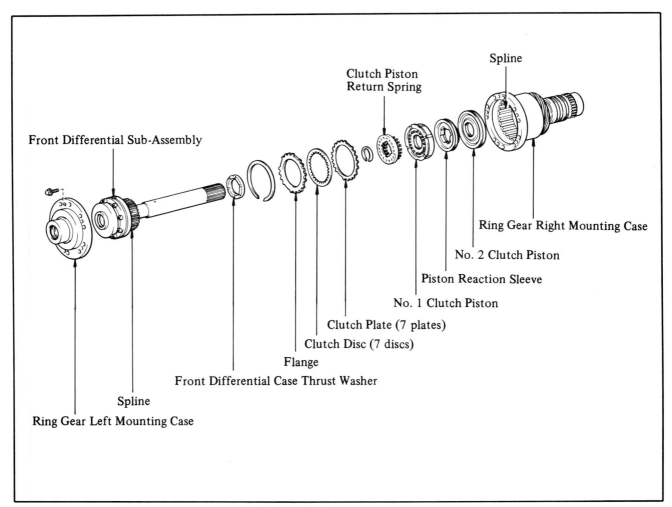

Inside the multiplate center differential are multiple discs, splined alternately to the inner shaft and outer housing. Direct contact between the discs transmits torque. *Toyota*

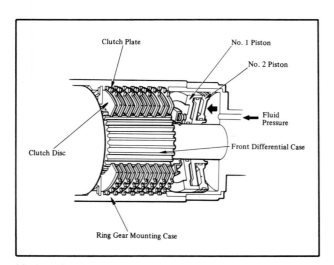

Hydraulic pistons, pushed by fluid pressure, force friction plates into contact to transmit torque. The same hydraulic principle operates the brakes on most vehicles. *Toyota*

therefore, feature three differentials: one each for the front, rear and center of the driveline.

Center differentials aren't necessarily located in the center of the car, just somewhere between the axles. An open center differential creates the same problems from front to rear as an open front or rear differential creates from side to side. So nearly all cars with three differentials feature either a limited-slip center differential or some provision for locking the open center differential in slick conditions. The latter can be automatically or manually controlled.

Limited-slip center differentials can be viscous couplings or Torsens, but often a manufacturer will choose a multiplate clutch device instead. Usually controlled by electronic or hydraulic actuators, the multiplate clutch-type differential is made just the way it sounds. Alternating discs—similar to those in a viscous coupling—are splined to the input and output ends of the center differential. These discs, however, are designed like the clutch disc behind a

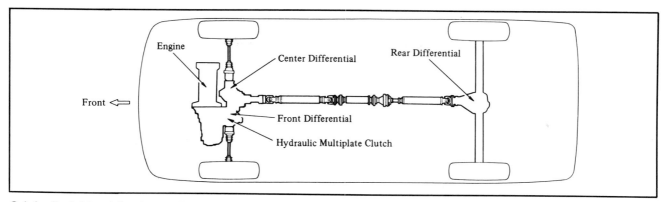

Originally laid out for front-wheel-drive, this car has been adapted to all-wheel-drive by the addition of a center differential, transfer case, driveshafts and a rear differential. *Toyota*

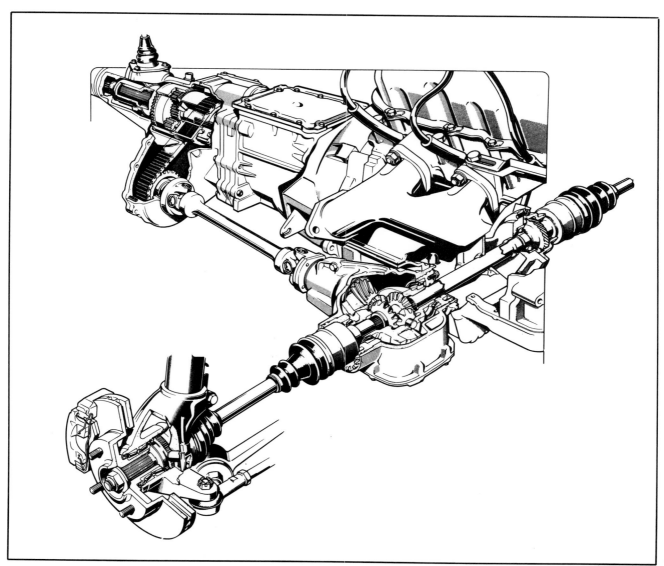

Originally designed for rear-wheel-drive, the Ford Sierra XR 4x4 uses a Morse chain to take power off at the transfer case. A separate front driveshaft takes power forward. An intermediate driveshaft to the left front wheel passes through the engine sump. *Ford of Europe*

manual transmission; they're made up of strong, simple friction material. When the actuator pushes the discs together, they provide lockup between the front and rear axles.

Front differentials are generally open, although some racing vehicles and show cars have experimented with controlled differentials on the steered wheels. A fully locked differential up front makes steering quite difficult, but locked rear and

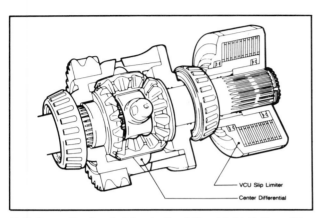

Full-time all-wheel-drive in the Mitsubishi Eclipse/Eagle Talon is provided by a viscous-controlled center differential in unit with a front differential, a transfer case and front driveshafts. *Mitsubishi of America*

center differentials guarantee at least three-wheel-drive under all circumstances—almost certainly enough to extricate a stuck vehicle.

Differences between the systems

Almost every all-wheel-drive performance car on the market began life as a two-wheel-drive platform that was later adapted to all-wheel-drive. This is probably the biggest determinant of how a car's particular drive system is laid out.

Cars originally designed for front-wheel-drive generally run a front and center differential in unit with the gearbox, and divide the torque between a basically standard front driveline and a shaft to the rear. Cars originally designed for rear-wheel-drive usually follow the principles of the Ferguson Formula, with a center differential behind the transmission and a regular rear-wheel driveline. At the front of the center differential, however, power is taken off by a Morse chain or gears and fed to a front driveshaft. A separate differential is located at the front of the car, and axle shafts go out from it to power the front wheels. Depending on the vehicle, the front axle shafts can go either under, in front of or through the engine's oil sump.

There are endless variations on this basic theme. The Audi V8, for example, has a rear-mounted transfer case and external front driveshaft, just as cars following the Ferguson Formula

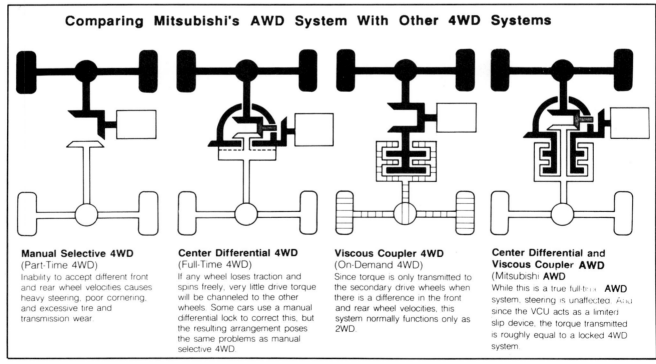

To help explain the full-time all-wheel-drive Eclipse, Mitsubishi drew up an excellent guide to its own and competing systems. *Mitsubishi of America*

layout have. Since the Audi's engine is ahead of the front axle, though, the driveshaft is very short and goes only as far as the front differential unit on the opposite side of the transmission.

Depending on the car's layout, intended use and intended price, its all-wheel-drive system can be simple or complex. Whatever the design of its driveline, however, it will fall into one of three categories: full-time, on-demand or part-time.

Full-time

Full-time all-wheel-drive is the most expensive and advanced drivetrain layout available. It's permanently engaged in all-wheel-drive, so the driver doesn't have to decide whether conditions demand all-wheel-drive or not—it's there regardless. When run with a viscous coupling, a Torsen differential or automatically locking differentials, this drive system is completely invisible to the driver. When it is run with manually locking center or rear differentials, the driver has to make a driveline decision only in the worst conditions.

The Audi 90 makes use of a Gleason Torsen center differential behind the front axle. ABS remains engaged at all times, and torque redistribution during wheel spin is immediate. *Audi of America*

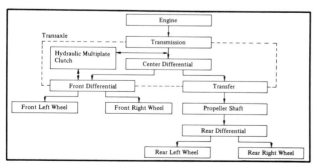

Torque flow through full-time all-wheel-drive changes with conditions. The key is variable flow from the center differential control unit, here a hydraulic multiplate clutch. *Toyota*

On-demand

Like the full-time system, an on-demand system can be left in all-wheel-drive 365 days a year without harming the machinery, thanks to a center differential. The benefit here is that wear and tear on the secondary driveline can be reduced by shifting out of all-wheel-drive and into two-wheel-drive for regular use. The system doesn't need to be as sophisticated as the full-time variety since the driver only has to live with it in poor conditions, so development and production costs can be kept down.

This system has a drawback as well: Often it isn't smooth enough to be used all the time, and a system that's disengaged won't do any good in emergency maneuvers or fast cornering. Another drawback is simply that the driver has to make the decision when to engage all-wheel-drive.

An alternative is automatic on-demand all-wheel-drive. As the name suggests, this system engages itself as soon as it senses the need for added traction. Until then just one axle is doing all or most of the work. One example is the viscous transmission, as employed in the Volkswagen Syncro models and Rally Golf.

With the viscous transmission, the viscous coupling doesn't merely oversee a mechanical transfer system, it *is* the transfer system. When the

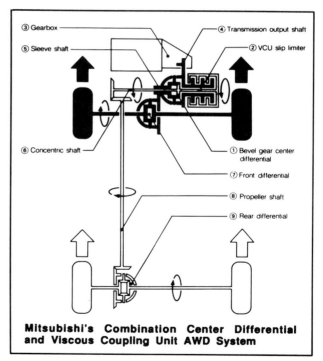

Mitsubishi's Combination Center Differential and Viscous Coupling Unit AWD System

Engine torque travels to the viscous-coupling-controlled center differential, then to the front and rear differentials along different paths. Mitsubishi's system is an addition to an existing front-wheel-drive platform. *Mitsubishi of America*

automobile is traveling down the road normally, axle speeds are relatively equal and almost no torque is transmitted through the coupling. If the primary drive axle starts to slip, however, the speed differential between axles activates the viscous transmission and torque is quickly transmitted to the secondary axle. This can happen in as

little as one-quarter rotation of the primary drive wheels. As soon as the axles begin turning at the same rate again, the viscous transmission becomes inactive once more.

Part-time

Lacking center differentials, part-time all-wheel-drives are the simplest and least expensive systems on the market. Part-time means just that: the all-wheel-drive system can be used only part of the time, because axle wind-up on dry surfaces would cause considerable problems.

A part-time setup requires constant attention from the driver since the secondary axle must be engaged and disengaged at just the right times. The system must be engaged as soon as slick conditions are encountered to avoid getting the car stuck, but it must be disengaged the moment the road clears up to prevent axle wind-up. Part-time systems are

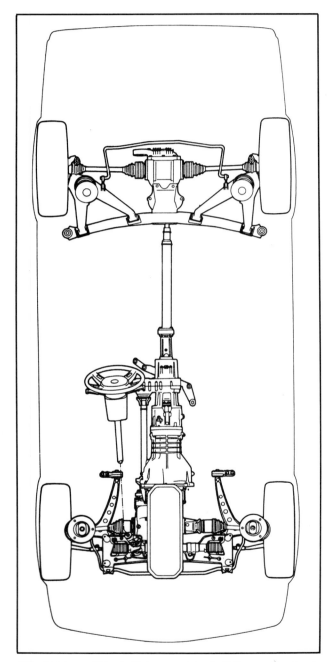

FF viscous differentials provide full-time all-wheel-drive on the BMW 325iX. Originally a rear-wheel-drive, this car added differentials, a transfer case and front driveshafts to go all-wheel-drive. *BMW of North America*

Volkswagen's Vanagon Syncro was the first to use viscous transmission, automatic, on-demand all-wheel-drive. Under normal conditions, only about five percent of the engine's torque goes to the front axle. *Volkswagen of America*

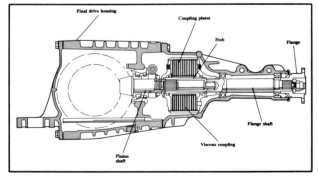

The inside of a viscous transmission unit seems almost empty. When the primary axle slips, a speed differential between axles occurs and the VT locks up. *Volkswagen of America*

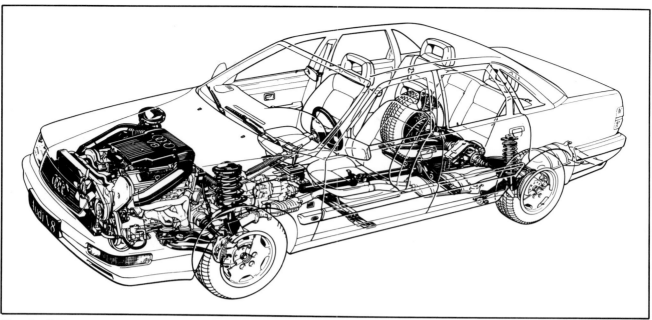

The Audi V8 sedan's engine rests ahead of the front axle but is not transversely mounted. A short driveshaft on the right of the transfer case takes power to the front differential. *Audi of America*

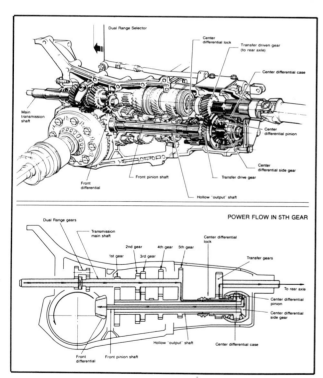

As with the Audi V8 sedan, the Subaru Loyale Sedan's longitudinal forward engine means a stubby, longitudinal front driveshaft is used. *Subaru of America*

Borg-Warner Automotive developed this compact transfer case, transmission unit for simple conversions from rear-wheel-drive to all-wheel-drive. *Borg-Warner*

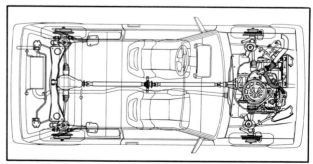

Subaru's Justy is a low-cost, pocket-sized, on-demand all-wheel-drive. Cost precludes much sophistication, but the system works great in heavy going. *Subaru of America*

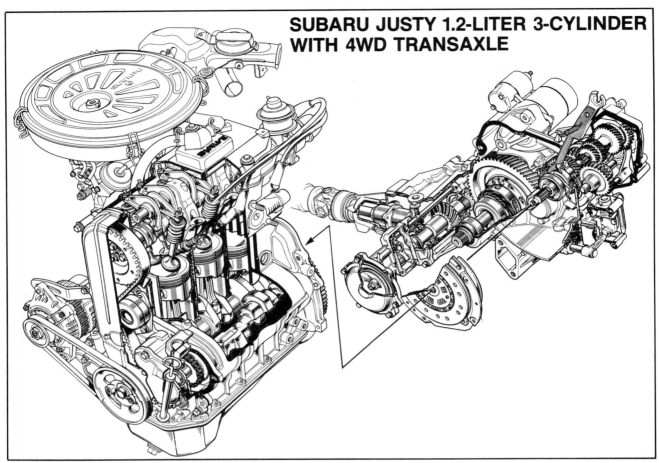

SUBARU JUSTY 1.2-LITER 3-CYLINDER WITH 4WD TRANSAXLE

The straightforward Justy driveline uses a bevel gear to turn the power 90 degrees, vacuum actuation to engage and disengage all-wheel-drive. *Subaru of America*

The Justy is the least expensive all-wheel-drive car sold in America. Its light weight, narrow tires and low cost make it a favorite with skiers and other snowbound types. *Subaru of America*

A street-bound homologation version of the RS200 put mid-engine turbo power, driver-selected all-wheel-drive and fantastic handling into the hands of some lucky Europeans. *Ford of Europe*

generally reserved for the least-expensive vehicles on the all-wheel-drive market.

Torque split systems

Another variable in all-wheel-drive systems is the torque split, the ratio of engine torque being sent to the front and rear wheels. Generally this is fixed at a constant rate. The action of the center differential can vary the ratio momentarily, but intentionally variable systems are also known.

The most sophisticated variable system lives in the Porsche 959, whose 450 bhp engine and rally car intentions justify its driveline's cost and complexity. In place of traditional differential, the 959 uses two ten-disc wet clutches under computer control to mete torque out to the axles as needed. The driver can select one of four modes based on his or her own impressions of the driving conditions, and a computer then adjusts the torque split

A stubby lever next to the gearshift of a Ford RS200 allows the driver to select torque distribution that is appropriate to road conditions and speed. *Ford of Europe*

A 50 percent front, 50 percent rear torque split gives the Subaru XT6 all-wheel-drive the same basic handling as its less-costly front-wheel-drive sibling's. Roadholding is dramatically improved, however. *Subaru of America*

A rear-biased torque split allows drivers of the European Ford Sierra XR 4x4 to kick the tail out with liberal applications of throttle. The Sierra was sold as the Merkur XR4TRi in America, but all-wheel-drive wasn't offered. *Ford of Europe*

The Peugeot 405 Mi 16X4 offers an optional Torsen rear differential. An all-wheel-drive version of the 405 was scheduled to reach American soil in 1991. *Peugeot of America*

by reading the driver's selected setting, wheel speeds, engine speed, throttle position and so forth. Secondary considerations have been programmed in to circumvent this loop system and provide special biasing under braking, low-speed use and other conditions.

Such sophisticated systems are rare, to say the least. Another variable system is the already-mentioned viscous transmission. Here, again, torque is automatically fed to the secondary drive axle when the need arises, but the mechanics behind the process are purely physical and no computer control is used.

Finally, high-performance all-wheel-drives like the specials built for World Rally Championship competition generally have a knob or lever near the gearshift that allows the driver to adjust front-to-rear torque split much as he or she would adjust the brake bias on a regular racing car. The driver

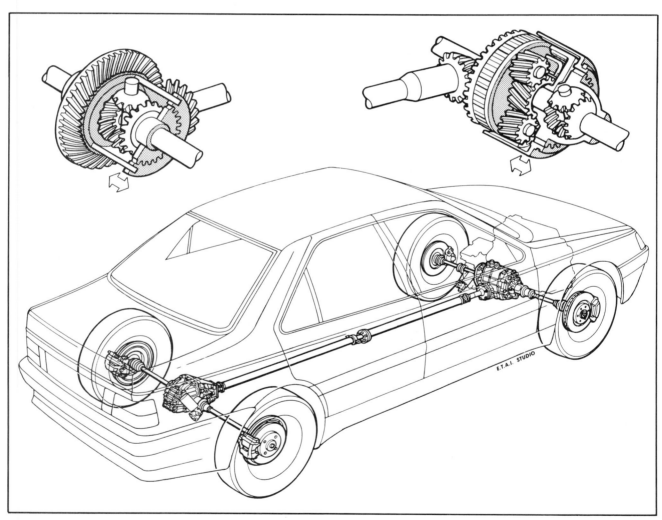

The planetary center and open rear differentials can be manually locked in the full-time all-wheel-drive Peugeot 405 Mi 16X4. *Peugeot of America*

generally dials in more rear bias for power over-steer at low speeds, more front bias for straight-line stability.

Most commonly, however, torque is split at a given rate from front to rear, and only the temporary action of the differentials can override this ratio. The two most common ratios of permanent front-to-rear torque split are fifty percent front to fifty percent rear and about thirty-four percent front to sixty-six percent rear. In fifty-fifty systems, generally used on platforms originally laid out for a front-wheel-drive, the vehicle takes on the power understeer characteristics of a front-wheel-drive car. With platforms originally designed for rear-wheel-drive, a rear torque bias is most often seen and power oversteer can be expected.

Drivers who are used to front-drive performance cars like Dodge Shelbys and Hondas will feel more comfortable in a car with a fifty-fifty torque split, whereas the reverse is true for previous Mustang or Corvette owners. Relearning the different handling characteristics of a front- or rear-drive automobile isn't difficult, but it takes time to get comfortable enough to count on yourself in an emergency.

Inside a few more drivelines

Vehicles with fixed torque splits need not be unsophisticated. One of the most interesting fixed all-wheel-drive systems comes from Mercedes-Benz.

The engagement and behavior of the Mercedes Dynamic Driving System is controlled at three levels. Speed sensors at each wheel and on the pro-peller shaft feed information to a central computer about the traction levels at all four corners of the car. The computer then decides whether to electronically lock the rear differential, momentarily throttle back the engine, or engage the car's 4Matic all-wheel-drive and send thirty-five percent of the torque to the front of the car.

Should the last alternative be chosen, 4Matic hooks up in three further stages. First, the front axle driveline is engaged. Second, if wheel spin continues the center differential is locked. Finally, if even more traction is needed the rear differential is locked.

Mercedes claims that exhaustive testing and development have made the system completely smooth and unobtrusive, despite the great number of driveline changes that are constantly occurring. The only way for the driver to tell what's going on is by status lights on the dashboard, not by the seat of his or her pants.

At the opposite end of the spectrum are the simple part-time systems found on low-cost offerings like the Ford Tempo all-wheel-drive. In the Tempo, the driver toggles an electric switch to simply throw the secondary axle in and out. The driver either has fixed-split all-wheel-drive or doesn't have it.

As with any other part of a car, the different all-wheel-drive systems all have their own advantages. Mostly these are measured on a sliding scale of usefulness against price. As with the car itself, it doesn't make much sense to pay for more than you need or at least want.

Chapter 3

All-Wheel-Drive Performance History

Who invented all-wheel-drive? Lots of people. The fact is, the all-wheel-drive has so many main characters there's no way to give just one or two most of the credit.

But while all-wheel-drive is almost as old as the automobile itself, until recently only a handful of people recognized its true potential. This minority believed that putting power to all four wheels of a car made good, common sense—and made for a faster, safer machine.

Pioneers like Joseph Laviolette, Otto Zachow, Harry Miller and the gang at Harry Ferguson Research helped to get all-wheel-drive developed

The 1903 Spyker can lay claim to being the world's first all-wheel-drive internal-combustion automobile. Joseph Laviolette and the Spijker Brothers were about eighty years too early with their ideas. Underneath the almost quaint bodywork of the Spyker was a downright modern chassis. All-wheel-drive complemented an 8.8 liter six-cylinder engine and a dual-range manual gearbox. *Jan P. Norbye*

and ultimately accepted. These pioneers deserve as much respect for the strength of their convictions as for the material accomplishments they left behind. Equal credit should go to today's more anonymous engineers, the ones who were given the task of turning the earlier ideas into honest hardware.

It's the pioneers, though, who make history.

Early efforts

The development of all-wheel-drive is the story of endless false starts, but eventually it all paid off. As best we can tell, construction of the first all-wheel-drive internal-combustion vehicle began in 1902. Fittingly, it was a racing car.

The beast came from Trompenburg, Holland, and the Spijker brothers who lived there. Along with their chief engineer, Joseph Laviolette, they built a surprisingly advanced machine incorporating an 8.8 liter inline six-cylinder all-wheel-drive car with three differentials and a three-speed, dual-range gearbox for six forward ratios.

Their Spyker automobile was more a technical exercise than anything else, but it was called a race car and it made a fairly good one. Within a few years, it traveled to England where it made a decent splash by winning the Birmingham Motor Club Hill Climb outright.

Shortly after the Spyker raised curiosity in Europe, but little in the way of cash for production, blacksmith Otto Zachow hit on all-wheel-drive for his own reasons. Zachow had paid good money for a slick new Reo, but he spent half of the time getting it stuck in mud and the rest of the time stuck in ditches. Zachow lived in Clintonville, Wisconsin, and the muddy tracks that passed for roads there ate the Reo alive.

One day Zachow discovered that he could back the Reo out of many ditches that he couldn't drive it from head first. A tinkerer by disposition—and probably sensing that there wasn't much future in blacksmithing anyway—he decided to make a vehicle with drive to all four wheels. What he came up with was a steam-driven all-wheel-drive truck, and it worked like a champ.

That first machine proved itself superior to every other conveyance in town during Wisconsin's long winters. It fought fires, rescued horses and generally made Zachow out to be one heck of a clever guy—which, of course, he was. It lacked a center differential, but as the problem had been Wisconsin's friction-free roadways in the first place, there wasn't much chance of axle wind-up being a problem.

After Otto had proven his truck's worth time and time again, his friends and neighbors decided the whole town should go into the automobile business and make copies of Zachow's vehicle. Shares were sold, and the Four Wheel Drive Auto Company, or FWD Co., came into being.

The high point of FWD's history probably came when the US Army staged a rally between Washington and Indianapolis to see if cars and trucks could compete with mules. The rally showed that not only could FWD's Battleship go anywhere mules went and then some, no other car even came close.

The Army wanted trucks rather than cars, of course, so FWD turned in that direction. By the end of World War I, 35,000 FWD trucks had been delivered and a competing machine, the Jefferey Quad, saw 30,000 made. The FWD name still appears on heavy trucks today.

The racers of 1932

Four-wheel-drive remained a part—a small part—of the truck world until 1932, when by coincidence three racing cars appeared with power at all their wheels. The most famous, at least in America, was a Harry Miller creation commissioned by our old friends at FWD. It was ostensibly built to allow study of the dynamics of all-wheel-drive at high speed, but the element of publicity was not to be overlooked.

A pair of FWD-Millers, one with a narrow-angle V-8 and one with an inline four, appeared in 1932. The four-cylinder car was the faster of the two, touching 175 mph with Mauri Rose's foot planted. Rose eventually got it to fourth place at Indy in 1936, but the FWD-Miller's performance was far from earth shattering. It was merely competitive— and for just competitive, plenty of two-wheel-drive cars were a whole lot easier to build.

But Miller, already the expert on both front- and rear-wheel-drive racers at Indy, liked what he'd found. He knew all too well the benefits and problems of both two-wheel-drive layouts, and he wondered if he couldn't rig up all-wheel-drive to be better than either of them. The idea intrigued him

Mauri Rose and the Miller FWD Special soldiered on to fourth at the 1936 Indy 500. Only Bobby Unser has placed an all-wheel-drive Indy car higher, finishing third in 1969. *Indianapolis Motor Speedway Museum*

enough that his involvement didn't end with FWD's departure.

At Indianapolis, the circumstances that really favored all-wheel-drive didn't exist. The cars spent most of their time at speed and a little wheel spin getting there manifested no great crisis. But there was the all-important matter of powering out of those four big turns. That was the secret as Miller saw it. All-wheel-drive let the driver get on the gas sooner, and harder, than did two-wheel-drive, not because of the added traction at the drive wheels but because the car was less sensitive to brake- and throttle-induced handling changes.

That was the theory, anyway. At Indy, however, theories are cheaper than grandstand hotdogs until they've been proven by checkered flags.

Le Patron and the Type 53

Europe's version of Harry Miller was Ettore Bugatti, and Bugatti just happened to be playing with all-wheel-drive himself in 1932. His most recent creation, the Type 51 Grand Prix car, was spinning its rear wheels instead of biting like it was

supposed to—no big surprise, considering its small size and 200 hp straight eight.

But unlike Miller's cars at Indy, Bugattis under the conditions of Grand Prix racing had simple, straightforward traction problems. The Grand Prix courses were dirty and loose, full of tight corners and riddled with elevation changes. Here all-wheel-drive might make a tremendous difference, assuming that the technical problems didn't prove too great.

The Bugatti Type 53, introduced the same year as the FWD-Miller, was a hardware lover's dream. In the middle of the chassis was a single aluminum case housing the transmission, transfer case and center differential. It took power from an input shaft off the crank, ran it through a bushel basket of gears, and fed it to offset final drives on the front and rear axles. The Type 53 also featured Bugatti's first independent front end, a funny-looking number with two transverse leaf springs acting as control arms and a chassis-mounted differential.

Though not a world beater on regular surfaces, the all-wheel-drive Bugatti was the best hill climber on the Continent. (The surface conditions of hill climbs were even worse than those of traditional Grands Prix).

Despite all the time and money spent in its creation, just one example of the Type 53 was ever

Ettore Bugatti's foray into the world of all-wheel-drive gave the world the Type 53. More successful as a hill-climber than a Grand Prix racer, the 53 now lives in France's Schlumpf Museum. *Jan P. Norbye*

The Type 53's heavy-looking front suspension, with upper and lower transverse leaf springs and friction shocks, worked well on rough surfaces. As usual, Bugatti passed over hydraulic brakes for cable operation. *Jan P. Norbye*

built. Though there's little word from those who drove it, it was most likely a handful at speed; that may explain how Jean Bugatti wiped it out at Shelsley Walsh—where it held an unofficial record—in 1939. In 1989, the restored chassis existed in the French Schlumpf collection.

The fastest all-wheel-drive on earth

The third appearance of an all-wheel-drive race car in 1932 came from England, in the form of a land speed record chaser from Reid A. Railton. John Cobb did the trick with the Napier-Railton and became, for a time, the fastest man on earth.

The behemoth racer probably had all-wheel-drive more out of default than by design. Two Napier V–12s powered the car, and it was simply easier to use one engine for each axle than to try gearing them together.

Other twin-engined all-wheel-drives, while rare, aren't unheard of: Robert Waddy had the Fuzzi in the thirties, Lou Fageol tried the idea with some Indy racers, and a variety of serious drag racers and exhibition cars combined all-wheel-drive with multiple American V–8s. Twin-engined experimentals have also been seen from the British Motor Corporation (the Twinni Mini Cooper that crashed spectacularly with Cooper himself on board), Volkswagen (rocketship Golf prototypes for Pikes Peak) and other established companies.

Even *Car and Driver* built a push-pull Honda CRX for the edification of its readers and its own curiosity. An entertaining exercise, it proved what many people have felt all along: Though the two-engine concept remains fascinating, it probably isn't the best way to go fast.

Twin-engined all-wheel-drives sound appealing since they don't involve all that mucking around with torque splits, driveshafts, center differentials and so on. What they do involve, though, is a polar moment of inertia the size of the USS *Iowa*'s and lots more weight than a regular all-wheel-drive with proportional horsepower.

Miller tries again

Harry Miller decided to give all-wheel-drive another whirl in 1938, this time with backing from Gulf Oil. His creation was as visionary for its day as the Spyker had been thirty-six years earlier.

Miller's trip to the Vanderbilt Cup races in 1937 probably had some bearing on the car's design. Like the Auto Union Grand Prix cars he saw there, his next all-wheel-drive racer had a midship-mounted engine and teardrop bodywork. There the similarities ended. Miller's car got its power from an inline six tilted 45 degrees from vertical and featured all-wheel-drive.

The Gulf-Miller, which on paper sounds more or less like a competitive race car for 1990, should have done much better than it did. Its best finish at the Brickyard was a tenth place in 1939. Neverthe-

The Type 53 body looked more like a freight train than a traditional sleek Bugatti. An imposing car to look at, the Type 53 made up in brawn what it lacked in beauty. *Jan P. Norbye*

less, George Barringer set a number of International Class D speed records in the car, taking the title for every distance from five kilometers to 500 miles. But all-wheel-drive probably had little to do with those records, except, perhaps, in getting the car up to trap speed quickly.

The Jeep

During World War II, not much happened on the all-wheel-drive front with the very large excep-

The go-anywhere World War II Jeep did just that. After the war, the Jeep gave birth to a whole new kind of vehicle, the off-road and sport-utility segment. *Chrysler*

The earliest Jeeps from American Bantam were smaller and had more complex bodywork than later Willys and Ford designs. American Bantam lost the bulk of the Jeep contract for lack of production facilities. *Chrysler*

tion of a rather unimpressive looking number called the Jeep and the other vehicles it inspired. The little cruiser from American Bantam, Willys and Ford gave birth to the huge off-road four-wheel-drive market that still outsells all-wheel-drive performance cars by a healthy margin.

Unlike the Millers and others, the Jeep shared the go-anywhere, loose-ground mission of FWD's trucks in lieu of pavement prowess. Therefore, the drive system was a step backward, at least in terms of technology, from what had come before on racing cars. Lacking a center differential, the Jeep had to be manually shifted out of four-wheel-drive on hard surfaces, a situation that still exists on many trucks and the remaining handful of part-time all-wheel-drive cars. The Jeep was perfect for its intended use—that use just had little to do with speed or finesse.

Jeep pieces did manage to find their way onto a number of homebuilt competition cars, in particular the AJB of Archie Butterworth. Butterworth's Steyr-engined hill climber was an impressive, if cobbled-up, racer that eventually fell into the hands of William Milliken, Jr., a slide rule boy from Cornell Aeronautical Labs. Milliken used the AJB to conduct some ground-breaking studies on all-wheel-drive vehicles, and for a bit of racing to boot. The AJB (also known as Butterball) now rests quietly at FWD's small museum in Clintonville, Wisconsin.

Postwar projects

After World War II, it took surprisingly little time for the world's auto enthusiasts to get back to work. In all-wheel-drive circles, one school quickly embraced the Jeep and took off in the direction of off-road, back country fun—the direction that bore the most fruit for the next thirty years. Another school continued applying all-wheel-drive to racing cars.

In the years between Bugatti's Type 53 and the invasion of Poland, Grand Prix racing had undergone a titanic horsepower war. German teams

The Porsche 360, commissioned by Cisitalia's Piero Dusio, might have been turned into a winner with proper development. The mid-engine layout was more than a decade ahead of its time. *Road & Track*

funded by Hitler had advanced the state of engine design to the point that wheel spin was becoming an almost crippling problem.

Prewar Grand Prix cars from Mercedes and Auto Union were trying to get more than 500 hp down to a greasy track through two skinny tires, and the result was spinning rear wheels at 150 mph. When the war came along, Germany's engineers had bigger things to worry about than racing cars and the problem was not resolved.

But the period left an impression on Ferdinand Porsche, the independent engineer who'd conceived Auto Union's sensational mid-engined Grand Prix racing cars. Released from French custody after the war, Dr. Porsche was most likely thinking about his next meal when Piero Dusio approached him with a proposition to design another Formula car.

Dusio, the Turin, Italy, businessman behind Cisitalia automobiles, wanted Porsche to do for Cisitalia what he had done for Auto Union. It was not important that Dusio probably hadn't the research and development funds to debug a toas-

The Porsche 360's supercharged flat twelve displaced 1.5 liters and produced about 300 bhp. A long finned case behind the engine housed the car's transmission, rear differential and transfer case. *Road & Track*

ter, as Porsche wasn't in the greatest financial shape himself.

Porsche's answer, the Type 360, was a Grand Prix car of a different stripe indeed. Like the prewar Auto Union, it featured a mid-mounted engine with a rear-mounted gearbox—this time a supercharged, 1.5 liter flat twelve mated to a five-speed transmission. Unlike the two-wheel-drive Auto Unions, however, the Type 360 had a long shaft that traveled under the engine to a front differential, and drive to the front wheels could be engaged and disengaged at will by means of a dog clutch.

Though the design was intriguing, the cash simply wasn't there to see it through. The 360 eventually wound up in Argentina, where it raced just once in 1953 and in two-wheel-drive only. The car resided in the Briggs Cunningham Museum in Costa Mesa, California, until the collection was liquidated in the late eighties.

Postwar calm in Grand Prix racing

The all-wheel-drive system of the Cisitalia might not have been necessary, since the car had only about 300 hp to contend with. Engine output had been throttled back with new regulations after World War II, and lower outputs coupled with better tires meant wheel spin was not a big problem in the first years of the new Formula One.

But when Mercedes-Benz decided to re-enter the Grand Prix arena it remembered well the traction problems of days gone by—and it had a hellaciously powerful engine up its sleeve. Therefore, the company's W.196, which went on to dominate Formula One, was first laid out with all-wheel-drive.

Late in the program, the Mercedes team learned it could get all the performance it wanted without driven front wheels, so that's what it chose to do. Mercedes correctly reasoned that as long as it won, why should it spend time and effort on the unknowns of all-wheel-drive?

Throughout the fifties, the situation remained that way. Racers like the BRM V-16 arrived with tremendous power, but tires and suspensions were being improved in kind. While all-wheel-drive would probably have helped them greatly, there were other, better-defined fixes to be tried first. If someone else could test, debug and prove the worth of all-wheel-drive, then other racing teams might well have embraced the technology. But no one wanted to be the guinea pig. Well, almost no one.

Dixon/Rolt Developments

Freddie Dixon, Harry Ferguson and A. R. "Tony" Rolt shared a common interest in all-wheel-drive, but they saw it as a stepping stone toward three different goals. Dixon, a hard-drinking racer-tuner from northern England, wanted to build an all-wheel-drive record car like the Napier-Railton. His Dixon Dart, however, would have to wait until someone came along with the proper backing; Dixon was a better racer than financier.

Rolt was Dixon's friend and employer—he paid Dixon to maintain a 1.5 liter English Racing Automobile (ERA) racer for him to play with. When one day they discussed the Dart, the younger Rolt suggested that Dixon build an all-wheel-drive road racer instead. That was more to Rolt's tastes and exciting enough that he would be willing to donate his ERA's engine to the project. The two formed Dixon/Rolt Developments in 1939 to stir the waters of all-wheel-drive.

Meanwhile, Ferguson—an Irishman with a fortune made in tractors and farm machinery—had four-wheeled ambitions of his own. Unlike many inventors, Ferguson was a good enough businessman to make tinkering worth his time, and he was always ready to try something new. What he eventually wanted to try was leaving the farm-bound machinery that had made him his money and getting into the auto business.

Far ahead of his time, Harry Ferguson dreamed of building cars that were safe. While Ralph Nader was still crash testing Tonka trucks in a sandbox, Ferguson was planning out the hardware that would save people's lives. An integral part of such a car would be all-wheel-drive. He'd seen how many skids, slides and other surprises it prevented with tractors, and he knew that it would make an even bigger difference on the road.

On top of his more agricultural interests, Ferguson owned a garage near the Ards Tourist Trophy circuit in Ireland, one of Dixon's favorite races. When racing the Tourist Trophy, Dixon prepped out of Ferguson's garage; it was simple, then, that two such inveterate tinkerers would become friends. They often discussed their mutual interest in all-wheel-drive.

When World War II came in 1939, Dixon/Rolt Developments, still without Ferguson, tried to sell the British Army on the concept of all-wheel-drive military vehicles. The War Office, however, had other things on its mind.

Without a prototype to show the government, Rolt submitted himself to the Army instead. He was taken prisoner at Dunkirk, France, and spent the rest of the war trying to get home again. He made it as far as the Swiss border on one occasion, and helped construct the famous Colditz Castle glider on another. Unfortunately, however inventive, he wound up in Germany for the duration.

Presumably Rolt had time to think of more than escape, and soon after the war Dixon/Rolt Developments had a Riley-engined all-wheel-drive prototype up and running. The single-backbone platform bristled with innovation: in addition to all-wheel-drive, it boasted four-wheel-steering and a single, centrally mounted disc brake on the main driveshaft. Power was delivered to front and rear by a main driveshaft in the single-backbone tube, which in turn drove worm-and-wheel final drives. Halfshafts with U-joints brought power out to the wheels themselves.

But despite its originality, the Crab lacked a few things in application. Steering, for instance, was accomplished by a series of links that brought the inside wheels closer together in a turn while pushing the outside wheels farther apart. It was a sensible layout on the surface, but Dixon and Rolt soon found why nobody else did it that way. It oversteered tremendously, and the wheels locked straight ahead when the brakes were applied. It was, however, a start. Dixon/Rolt Developments had asked for nothing more.

Harry Ferguson Research

Ferguson, who'd moved to England after the war, one day rang up old friend and tenant Dixon. When Dixon mentioned that he'd made a start on an all-wheel-drive road car, Ferguson decided to have a look.

What the Irishman saw convinced him that Dixon and Rolt were serious. He offered cash to become a third partner, and the enterprise suddenly became a going concern.

The only thing left to do was find a professional engineer who could backstop the self-taught trio, and an all-wheel-drive production model would certainly follow. In any case, the Crab had taught Dixon and Rolt that such a person might be good to have around. The three found Claude Hill, an engineer of impeccable credentials, late of Aston Martin.

With all the new developments at Dixon/Rolt—namely a sudden influx of money and expertise—it was decided to toss out the old order and restart the company from scratch. The new name, come 1950, was Harry Ferguson Research Limited.

What would distinguish Harry Ferguson Research from all the organizations that had previously dabbled with all-wheel-drive was its dedication to one goal. Ostensibly the company existed to develop technologies that could be sold to the British motor industry, and it took on development contracts to supplement its budget. But in practice the group was an almost single-minded think tank for all-wheel-drive. While other groups saw all-wheel-drive as a means to an end, Harry Ferguson Research saw it as the end itself.

The next prototype was the R2. Carried over from the Crab (renamed R1 retroactively) were the worm-and-wheel final drive, backbone chassis and centrally mounted single-disc brake. The four-wheel-steering was abandoned as more trouble than it was worth.

Meanwhile, one of the team's money-generating outside contracts involved development work for Dunlop's disc brake program. The connection with Dunlop led to Ferguson's zealous adoption of

antilock braking twenty-five years before it, too, became the latest thing in safe automobiles.

Unfortunately, all this hustle and bustle began to grate on Freddie Dixon's nerves. What had started as an enjoyable, try-it-and-see sideline business with Tony Rolt had developed into a technical engineering shop with Claude Hill holding all the degrees. Dixon and Harry Ferguson Research parted company in 1952.

After the R2 came the R3, and with this machine most of the important work of Harry Ferguson Research began. To prevent axle wind-up and its associated mechanical nastiness, by now the team knew that some sort of differential between front and rear axles was necessary. They were up against the old bugbear of open differentials, though: Bad traction on a single wheel immediately reduced the vehicle from all-wheel-drive to one-wheel—the wrong one—drive.

Ferguson decreed that Harry Ferguson Research should build "a diff that should diff when it should diff and not diff when it should not diff." He left Claude Hill to figure out the details.

Hill did just that. He used gears on either side of an open center differential to drive free wheels controlled by clutch units. The clutches became active only when preset speed differences between the front and rear driveshafts occurred, and they effectively locked the two shafts together under those conditions.

It worked like a charm, but only when going forward. By the time provisions were made for reverse, the system became complicated indeed. It was incorporated into the R3 regardless, along with a torque converter, manual gearbox combination that allowed manual or automatic shifting, and an all-original horizontally opposed four-cylinder engine.

Aside from all-wheel-drive, the most impressive feature of the R3 was its antilock braking system (ABS). Dunlop had developed Maxaret, an ABS for aircraft, and Ferguson borrowed the technology for his road car.

The aircraft system was hopelessly complex, in that separate sensor units were needed for each wheel. With Ferguson's all-wheel-drive, however, it was thought that just one unit would work as well since all the wheels were effectively tied together. Antilock brakes quickly became the second leg of Ferguson's safety car.

When the company finally felt it was ready to make a roadworthy prototype, it created the fully enclosed R4, a Morris Minor look-alike with Hill's flat four engine hanging out front. Harry Ferguson took the well-detailed prototype out to show it off in much the same way he'd demonstrated his tractor equipment twenty-five years earlier. Despite a good deal of manufacturer interest and Ferguson's great flair for demonstrations, no auto maker took the bait. One close call was British Motor Corpora-

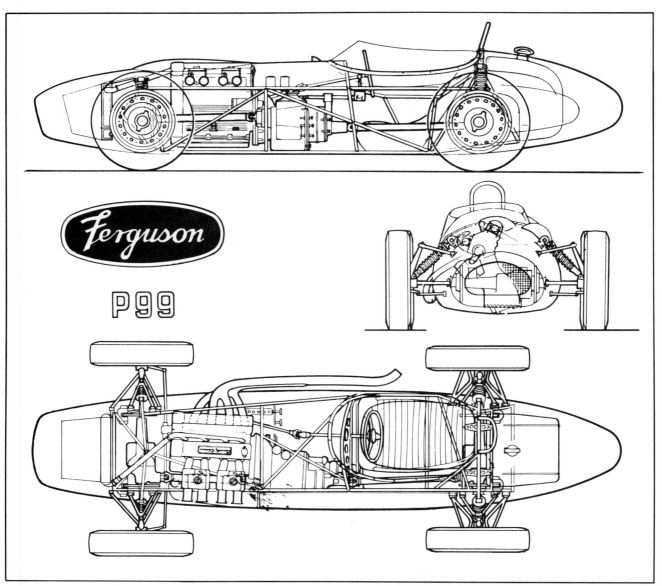

Aside from all-wheel-drive, the Ferguson P99 was a basically straightforward racing car. Its Coventry Climax four-cylinder was tilted to one side for a lower hoodline and to make room for the front driveshaft. *Road & Track*

tion (BMC), but when negotiations stalled that firm went ahead with the program that became the front-drive Mini.

By 1960, the technical problems of full-time all-wheel-drive had effectively been eliminated by Ferguson's crew. That turned out to be the easy part: even though it worked as promised, manufacturers were reluctant to invest in Ferguson's system. They simply doubted that buyers would pay for their own safety, and they were probably right. To tool up for Ferguson's devices meant more money than they were willing to gamble, and it didn't help that Ferguson had successfully sued Ford for patent infringement at an earlier date.

The decision was made, then, to demonstrate all-wheel-drive's effectiveness not just to manufacturers but to the public at large. Harry Ferguson knew how to put on a show, and he knew that what the company needed right then was not just a show but a spectacle.

There was no spectacle in Europe like Formula One.

The P99

In 1960, a massive campaign was launched to develop a Formula One racing car with Ferguson Formula components. The project went from start to finish within nine months, despite the tremendous amount of development involved and the death of Harry Ferguson in October of that year.

The best sources available were contacted for help: Rob Walker was asked to be team manager, Hardy Spicer supplied U-joint technology, Colotti was given the nod for the gearbox and next-door-neighbor Coventry Climax was picked to supply the team's engines.

In construction, the car was more or less like any other front-engined Grand Prix racer. Remember that rear-mounted mid-engines were the coming wave at the time. Accepting that all wheels were driven and that a Ferguson transfer box with Maxaret ABS sat behind the powerplant, the mechanical specifications were pretty straightforward: tube frame, independent double wishbones at all corners, five-speed transmission.

Stirling Moss was chosen as team driver, and by the Silverstone event in 1961 everything was ready—almost. The P99 had persistent braking troubles early on (oddly enough, the Maxaret unit was only coincidentally involved with them). The first problem come race day was a spongy pedal, later traced to faulty braking seals. But Moss had elected to drive his Lotus, so it was Jack Fairman who had to retire when a countershaft—over-stressed from forced downshifting—failed after three laps.

The next outing, Aintree, was important for Moss. All looked well when he set the fastest time in practice with the P99, but a caliper bridge-pipe failure suddenly made him wary once more of the Ferguson's braking reliability. This time, however, Moss' Lotus conked out in the race while Fairman and the P99 kept running strong. Then Fairman ran over a piece of debris and knocked loose an electrical lead, prompting a push from a corner worker and a trip to the pits. Moss took over the car and set some blistering times in pouring rain, but he was promptly black-flagged for Fairman's aided start. For Moss it was a bad day, but for Ferguson the results were fantastic. It seemed to have, at least in inclement weather, the fastest car in Formula One.

At the car's final 1961 outing, the non-points Oulton Park race, the P99 earned its first victory. This time Moss led going away and won in convincing fashion. Unfortunately, many spectators cited the lack of established competition in this postseason race as contributing to his win, and Ferguson was denied some of its due.

Moss later gave *Motor Trend* readers his opinions of the car. He stated that while the P99 didn't go on to great success in Grand Prix racing, its combination of all-wheel-drive and antilock brakes pointed the way to the future. Moss came away from the P99 episode convinced that Ferguson was onto something good.

The Ferguson Formula transfer case, Maxaret ABS and center differential unit of the mid-sixties would bolt right into many rear-wheel-drive cars. Harry Ferguson Research figured mass production of the Ferguson Formula cars would add about $500 to the vehicle's base price. *Road & Track*

The P99 was a fine-looking Formula One car, even with the odd cut-down plexiglass canopy. Unfortunately, the days of front-engined GP cars were quickly coming to an end. *Jan P. Norbye*

Extremely clean and compact, the P99 stuffed inboard disc brakes and a bushel of all-wheel-drive componen-try into its modest tube frame. The FF transfer case with Maxaret sits behind the oil reservoir. *Jan P. Norbye*

A five-door station wagon body reflected the practical market Harry Ferguson Research wanted to tap with the R5 prototype. Later all-wheel-drive proponents made their prototypes more racy and attention getting. *Road & Track*

Ferguson in the sixties

Ferguson's next and final full-scale prototype production machine was the R5, a sophisticated five-door wagon with all-wheel-drive, ABS, safety harnesses, fingertip controls (which were still unheard of in production automobiles), an impact-engineered engine compartment that absorbed energy in much the same way that modern Mercedes-Benz and Volvo automobiles do, and the expected Ferguson flat four with torque converter. It was well received by the press and public, but as usual there were no takers in the auto industry.

At the same time, however, two other projects came up that would have more immediate consequences. First, American Indy car owner Andy Granatelli wanted a Ferguson-equipped chassis to carry his Novi engine. Studebaker and STP sponsored the project, and Ferguson supplied basically a scaled-up version of the P99. Though the performance of the cars themselves wasn't remarkable, Granatelli and Ferguson would continue their association for as long as the United States Auto Club allowed all-wheel-drive at Indianapolis. (In

1964, BRM had also looked into all-wheel-drive, but nothing came of the project.)

The other big news of the mid-sixties came in the form of a Ferguson-equipped automobile hitting the roads at last. The manufacturer was Jensen, an exclusive sports-luxury constructor, and the automobile was originally to be an all-wheel-drive version of its Chrysler-powered C-V8. When Jensen switched to the newer Interceptor model, though, Ferguson Formula went with it.

The $13,300 Jensen FF was a far cry from the everyman's safety car Harry Ferguson had originally envisioned. Regardless, it gained acclaim as the most sure-footed supercar most testers had ever encountered and raised flagging spirits at Harry Ferguson Research considerably. Though just 316 Jensen FFs were sold before production ceased in 1971—and their glamour rubbed off more on Jensen's regular Interceptors than on Ferguson's hardware—they remain important as the first example of a true performance all-wheel-drive production car.

Harry Ferguson Research also tried to generate public interest with a number of company-funded conversions in the late sixties, including a particularly successful effort built on a Mustang

All-wheel-drive Mustangs that had been converted by Harry Ferguson Research negotiated icy test tracks that would have sent regular Mustangs sailing across the road. American and European manufacturers were impressed by the cars, but didn't sign on to build replicas. *Road & Track*

Andy Granatelli asked Harry Ferguson Research for a P99-esque Indy car to carry the powerful Novi V-8 engine. Jim McElreath and Studebaker sponsorship carried the project into the 1964 race. *Indianapolis Motor Speedway Museum*

Harry Ferguson Research built all-wheel-drive Mustangs with Ferguson Formula components to demonstrate the superiority of all-wheel-drive. Externally, the cars appeared completely stock. *Road & Track*

The STP Turbine car of 1967 was close to victory when a non-all-wheel-drive part failed and robbed Parnelli Jones of the win. Jones sat next to the long turbine engine. *Indianapolis Motor Speedway Museum*

coupe. But even though the Mustang had been given exemplary marks from manufacturers, government officials and safety organizations alike, the per-unit cost of $500 seemed too high for Ford. The company believed the average buyer would be inclined to save the extra money and wager his or her life in collateral.

GKN Birfield

Rolt, realizing that Harry Ferguson Research was pounding against a brick wall, decided to change strategy. He turned production rights of the Ferguson Formula over to GKN Birfield, a giant British parts and gear manufacturer, in the hope that it would be able to promote the technology more effectively. GKN went on to convert more cars with about the same results, finally coming up with the FFF 100—a one-off supercar that made Jensen's exclusive FF look like a milk float.

Motivated by a 600 hp Chrysler Hemi, the FFF 100 was good-looking, blindingly fast and remarkably well-mannered. Wonder of wonders, it led to even more good press for the Ferguson system and even less honest enthusiasm from the world's increasingly money-minded car markers.

The second coming of all-wheel-drive racing cars

In the racing world, however, all-wheel-drive was far from dead in the late sixties and early seventies. Granatelli continued to use it for the cars that superseded the Novi-driven Indy specials of 1964, 1965 and 1966. His new-for-1967 cars boasted honest-to-goodness turbine engines, and their tremendous power could only get to the ground through all four wheels. In 1967, Parnelli Jones had Indy wrapped up for Granatelli when a minor, non-all-wheel-drive component failed and put him out just three laps from the finish.

The next year, Lotus, too, joined the turbine, all-wheel-drive ranks. Again, minor failures took the cars out after they had dominated the better part of the race. Finally, when turbines were effectively banned by inlet restrictions in 1969, Bobby Unser took an Offenhauser-powered all-wheel-drive chassis to third for the highest-placed all-wheel-drive finish ever at Indy. The next year all-wheel-drive cars were banned from the Brickyard, and the whole experiment drew to a close.

All-wheel interest shifted across the ocean, however, when Cosworth's ubiquitous Formula One V-8 became so strong and peaky that racing teams doubted there was any other way to get its power to the track.

The first to get serious about all-wheel-drive for Formula One since the P99's time was Cosworth's own Keith Duckworth. Cosworth's only complete racing car was unsuccessful, but other teams noted the all-wheel-drive idea and approved of it. Soon all-wheel-drive interest was piqued everywhere in the sport. Lotus, McLaren and Matra all produced trackworthy all-wheel-drive Formula One cars, but success eluded each. In general, they were all plagued by power understeer and the driver's inability to cope with it. While attempts were made to proportion more torque to the rear wheels, tire and airfoil technology came nipping at all-wheel-drive's heels and was soon embraced in its stead.

One can't blame teams for going this simpler route, though it would have been fascinating to have seen a car with all-wheel-drive *and* advanced aerodynamic aids. We just might, however, get to see something like this in the prototype racing of the nineties.

Lotus' 1968 turbine car used a more compact power-plant than did Granatelli's machine and beautifully simple bodywork. All-wheel-drive seemed to be the only way to get a turbine's tremendous power to the track. *Indianapolis Motor Speedway Museum*

AMC and Subaru in the seventies

By the early seventies, all-wheel-drive was looking like a lost cause in production cars. Both Indianapolis and Formula One had left it behind, the former by rules and the latter by choice. The Jensen FF was gone, and its show car successor the GKN FFF 100 had aroused little manufacturer interest.

The technology of all-wheel-drive didn't remain at a standstill, at least, because there was always the off-road crowd to be satisfied. In one important development, Jeep introduced its Quadra-Trac system in 1973. Quadra-Trac used a center differ-ential with viscous limited slip to provide functional full-time all-wheel-drive. No more shifting in and out of four-wheel-drive depending on the road surface. Chrysler's New Process gear-making division introduced a similar system at almost the same time, this one a bit more advanced and based on Ferguson patents. The Rover group also took the viscous route for its independently sprung Range Rover, which still uses viscous couplings today.

But in 1975 a small and (at least in America) generally unregarded Japanese company took a chance on all-wheel-drive that made it an almost

An unassuming Subaru wagon introduced part-time all-wheel-drive to the masses in 1975. Pundits declared it would be a failure, but time proved them wrong. *Subaru of America*

49

overnight sensation. The funny-looking front-drive cars from Fuji Heavy Industries—better known as Subaru—became the first widely available all-wheel-drive automobiles in history.

Just like the first safety car Harry Ferguson had envisioned, the little Subarus were affordable enough for anyone who wanted one. Because they were so inexpensive, a center differential was out of the question, and they were only part-time all-wheel-drive cars. They made do with a respectable enough part-demand system, however, which was just fine for the people who bought them. These were mainly snowbound families, ski bums and folks who just wanted to be a little different.

For most of the year, the Subaru was just another Japanese economy car. But when the arctic winds blew, its all-wheel-drive, small size and narrow tires let it speed through muck that had everything else but Jeeps and Blazers stuck fast. The extra rear-drive wheels were inoperative in decent weather, so they didn't do a thing for performance in normal conditions. It's doubtful that overpowered front wheels were a big problem on early Subarus, though.

In 1979, a car came along as a 1980 model that was not much racier but still quite functional. AMC's Eagle—not to be confused with the later

Chrysler division of the same name—set out to beat Subaru at its own game. The first full-time all-wheel-drive automobile ever sold in any kind of quantity, like the Subaru it was much more foul-weather workhorse than performance car.

Essentially a standard AMC family car that was jacked up for ground clearance and fitted with all-wheel-drive, the Eagle nevertheless boasted an admirable viscous differential all-wheel-drive system. It was AMC's first sales success in years. Bigger than the Subaru and more familiar to people who usually bought American cars, the Eagle got a foothold in the same snowbelt market.

Car and Driver magazine, not usually given to AMC family sedans, saw potential in this car. It got hold of an Eagle with the idea of making it more sporty, perhaps because its Michigan-based offices spent a good part of the year covered in snow. Its Europeanizing job was somewhat less successful than it might have been, but the Alter Eagle was a look straight into the future of performance cars.

Enter Audi

In 1979, all-wheel-drive became legal for FISA's World Rally Championship. Within five short years, this relatively unregarded event would result in the full and final acceptance of all-wheel-drive, not just

Audi's sleek, understated Quattro Turbo Coupe ushered in the modern era of all-wheel-drive performance cars.

Magazines hailed it as the first of a new generation of supercars. *Audi of America*

An evolution model of the long-wheelbase Quattro, the A-2, continued Audi's winning streak in rallying. Bruno Kreibich campaigned one in SCCA PRO Rally events, but he couldn't beat top Audi man John Buffum to the title. *Audi of America*

in rally cars but in performance cars in general. By 1984, almost every manufacturer who didn't have an all-wheel-drive car would be working feverishly to get one. And the few who weren't looking actively were certainly keeping their eyes on the technology, for it was obvious that this time all-wheel-drive had come to stay.

The story of Audi's Quattro and the World Rally Championship is told in Chapter 4. In short, Audi was the only company that took up the all-wheel-drive challenge in 1979, and consequently was the only one that could win most rallies when all-wheel-drive turned out to be an almost unfair advantage.

Audi's engineers overcame some of the handling problems previously associated with high-performance all-wheel-drives, and its drivers learned to deal with the rest. Its Quattro Turbo Coupe proved that high-performance automobiles—*real* automobiles, not the lightweight oddities of Formula One and Indy—gained tremendous benefits from all-wheel-drive under all but the most pristine, high-friction conditions. Other cars from Porsche, Peugeot and Ford were left to prove that there were advantages to be gained even there.

What laurels the racing Quattro Turbo Coupe missed, its street-bound sibling picked up. Far from the coal-cart ride and shoddy construction of the usual homologation special, the Audi was a smooth and thoroughly refined peformance-luxury car. It was similar to the Jensen FF in mission and personality. Both were fast, but performed with such dignity that their speed seemed effortless.

But while the Quattro was definitely a car of the eighties—many called it the first supercar designed for the modern world—it lacked sophistication in one area: its driveline. Backing up its turbocharged five-cylinder engine was an all-wheel-drive system less sophisticated than the Jensen's Ferguson Formula pieces of fifteen years earlier. The problem was simply one of packaging: There wasn't room for any Ferguson Formula goodies inside the Quattro driveline because the car from which the Quattro was derived wasn't originally planned as an all-wheel-drive.

None of that mattered, however. In the end, the Quattro Turbo Coupe performed so well that buyers lined up for it. At $35,000 it was a good deal more expensive than the rival Porsche 911SC, but to many there was no comparison.

Besides putting Audi—which was doing pretty well anyway—into the same image league as BMW and Mercedes, the Quattro managed to once and for all implant the idea of all-wheel-drive into the minds of enthusiasts. And by convincing enthusiast drivers and the press, the word of all-wheel-drive came to the masses. Audi, on the wings of an

The requisite 200 production models of Peugeot's 205 Turbo 16 were lined up with project and factory personnel at the end of March 1984. Lucky French street drivers snapped up the cars that weren't destined for rally courses. *Peugeot of America*

experimental race car and a high-priced road car, accomplished in short order what Harry Ferguson and others weren't able to do in four decades.

Group B supercars

A pleasant offshoot of rallying's all-wheel-drive revolution was the 200 plus homologation specials that manufacturers had to build to make their own all-wheel-drive racers legal for the World Rally Championship. European buyers found themselves with an entirely new generation of supercars to choose from by the mid-eighties, all of them featuring all-wheel-drive and most with tremendous amounts of power. In addition to Audi providing the Quattros, Peugeot offered the 205 Turbo 16, Austin-Rover the MG 6R4, Ford the RS200, Porsche the 959, and so on.

All of these cars made great road machines in their own right, offering performance and in many cases considerable luxury. The collector value of, say, a pristine RS200 will no doubt be great in the years to come.

Offerings in the eighties

With all-wheel-drive suddenly an accepted—and for many the only—way to power fast cars, manufacturers scrambled to introduce their own all-wheel-drives throughout the eighties. The ones who were already working on racing cars—most notably Peugeot, Lancia and Ford—had the jump on everyone else, but other manufacturers dusted off the old information packets that Harry Ferguson Research, FF Developments and GKN had been sending out for years. They discovered that all-wheel-drive need not be a tremendously difficult conversion.

The most excitement came from glamour cars like the Porsche 959. Ferrari unveiled a concept car called the 408 with stressed aluminum construction and all-wheel-drive. (So far Ferrari still has no official all-wheel-drive intentions, but this company loves nothing better than to disinform nosy reporters.)

The rest of the eighties were spent with all-wheel-drive introductions following one after the other. A few appeared for the wealthy, then some for the not-so-wealthy and then more targeted somewhere in between. By the late eighties, there was an all-wheel-drive car for every market slot from economy car to supercar. For sedan lovers there was the BMW 325iX and Pontiac 6000 STE AWD. For boy racers, the Eagle Talon, Mitsubishi Eclipse twins and the Mazda 323 GTX offered superb all-wheel-drive systems coupled with low cost and double-overhead-cam turbocharged engines. For the sophisticated enthusiast the Porsche Carrera 4, a virtually all-new version of the 911 built around an all-wheel-drive platform, arrived in 1989.

Audi offered all-wheel-drive across the board, including the V8 sedan and a reiterated version of the Quattro Coupe. Toyota had All-Trac drive running in its sporty Celica, the family-minded Camry and even the economical Corolla. Minivans from Mazda and Nissan, station wagons from Dodge and Honda . . . the list goes on. And the selection will only keep getting better.

But the surprising thought going into the nineties wasn't that the market offered so many all-wheel-drives, in so many different classes. It was that we allowed ourselves to wait so long before this state of affairs came about.

Chapter 4

Rallying

In the United States, rallying is a largely misunderstood motorsport. Many people confuse PRO Rallying, the high-speed competition, with road rallying, where the object is to follow a given route at a uniform, usually low, speed.

The Sports Car Club of America (SCCA) says that "PRO Rallying is, quite simply, racing. In fact, many people refer to it as 'forest racing' or 'dirt-road racing.' Its only similarities to its road rally brother are that 1) cars compete one at a time on the road, receiving scores at timing controls; 2) each car includes a team of two people—a driver and a co-driver; and 3) a route book is used to assist in following the course during the event. Other than that—forget all your perceptions of what a rally might be!

"In the sport of PRO Rallying, modified production automobiles race over closed sections of public roads—and the competitor with the lowest accumulated time at the end of the event wins.

International rallying the way it *used* to be. The plebeian-looking Ford-Lotus Cortina Mark I featured a double-overhead-cam Lotus engine. It was one of the first high-output rally specials. *Ford of Europe*

All-wheel-drive and serious horsepower changed the face of international rallying. The Ford RS200 never got a chance to mature in the sport, but even early versions like this were plenty fast. *Ford of Europe*

Tackling the East African Safari, a Cortina gets ready to kick out its rear end. A rear-wheel-drive, front-engine layout made for great power slides. *Ford of Europe*

Rallyists compete over 10 to 20 'special stages' during the course of an event, with varying success, with the lowest final total again being declared the winner.

"PRO Rally cars are high-performance race cars that have been prepared not only for speed but also for safety and dependability. They, and their occupants, must be able to take the punishment of competing, throughout the night, on poorly-maintained dirt roads as well as smooth, paved ones. PRO Rally drivers are endurance athletes who must have the ability to go as fast as possible over roads they have never seen before, in conditions that range from forest rain and snow to desert heat."

PRO Rallies are held from February to December, a season even longer than the one SCORE/HDRA off-road racers endure. Cars race against the clock one at a time, and the events can be anywhere from ten miles long in a Coefficient 1 Divisional rally to more than 400 miles long in a major National event. In the past, everything from Volvos to Jeeps to Audis has been a class act of rallying.

Though the scores are made up primarily of cumulative times on the special stages, penalty points can also be assessed as the cars travel over regular roads at legal speed from one stage to the next. Penalties are issued for lateness *and* earliness, so competitors have to be on their toes. Too

The front wheels contributed little to the performance of this rally car. Engine tuning and a stripped-down body shell gave good speed regardless. *Ford of Europe*

Two cams and two Weber carburetors made the Lotus-Cortina's powerplant pretty sophisticated in the mid-sixties. Lack of specialization allowed a road race version in virtually identical form. *Ford of Europe*

Fording a stream with a two-wheel-drive rally car can be tricky. If the rear end bogged down on this Mark II Cortina, the driver's race could be over. *Ford of Europe*

Straightforward rally specials used simple tuning tricks for speed, so rally and road racing cars were little different. *Ford of Europe*

much time in transit can put a car over the Maximum Allowed Lateness and put the whole team out for good.

Most important, the drivers have never seen the route before, so they have no idea what sort of curve might be coming up next. They rely on years of experience and a good deal of nerve to decide just how fast they can go into the next blind corner. Therein lies the skill, and the thrill, of professional rallying.

Looking at the conditions in which some rallies are held, you'd almost think the organizers *try* to harry car, driver and codriver (navigator). Foul weather and dusty roads are no excuse to call a rally off, yet PRO Rally drivers say that their sport, for all its difficulty, is remarkably easy to get hooked on. (Later in this chapter we'll discuss the things you'll need to know if that happens to you.)

But rallying also provided race fans with one of the most exciting and technology-intensive periods in racing history: FISA's World Rally Championship Group B. Without it, you probably wouldn't be driving an all-wheel-drive now. And knowing a bit about Group B can also help you understand the state of rallying today.

Thunder cars of the World Rally Championship

Today's all-wheel-drive renaissance is a by-product of the great interest generated when Audi's Quattro took WRC Group B racing by storm. WRC rallying remains today what it was then: second only to Formula One as Europe's favorite motorsport. Events are held at all levels of competition, much as they are in America's NASCAR series. There's everything from local weekend club races to high-buck, high-tech, high-speed international events like the Tour de Corse and the Rallye Monte Carlo.

Similar to PRO Rallying events, WRC events are made up of stages run flat-out against the clock. If the thought of professional racing drivers charging full-speed over a snow-covered Alpine road sounds dicey, you can see why all-wheel-drive made such an impression when it finally arrived in force with the Quattro.

At first, rally cars were unmodified or barely modified production cars. Family sedans—Fords, Saabs, Sunbeams and Volvos—often ran as well as or better than sports cars like Triumphs, Alfas and Jaguars. What made a car fast on regular pavement—horsepower—didn't necessarily help on snow and gravel. In fact, more often than not lots of power was a handicap. Without the traction to get it down, too much power just spun the wheels.

The modifications on even works entries in the fifties and early sixties were usually restricted to safety equipment, lightweight accessories, extra lighting, and perhaps a bit of intake and exhaust tweaking. That all began to change in the mid-sixties when purpose-built rally specials arrived in the form of the Austin-Healey 3000, Austin Mini Cooper S and Ford-Lotus Cortina. Big advances in tire and suspension technology helped these cars get more power to the road, and their strong engines made the most of it. The arrival of sponsors in the early seventies brought even more money to the series, running technology (and costs) up further still.

The heart of the matter was that the WRC had grown up by 1980. Nevertheless, despite all the advances in technology, it was still pretty easy to get more power from an engine than even the best rally cars could put to the ground. The key to speed was still in better tires, stronger brakes, slicker suspensions and—perhaps most of all—tougher cars.

With a few notable exceptions, the hot rally cars at the end of the seventies were front-engined, rear-drive coupes based loosely on production models. They weighed a bit more than a ton and sported about 250 hp—which was more than they could use, but it made for great-looking power slides. A few sophisticated cars like the Lancia Stratos (which used a mid-mounted Ferrari V-6) ran well, but by and large the economy car silhouette was the way to go.

Arrival of Audi's Quattro Turbo Coupe

Since traction, or the lack of it, was the great equalizer in professional rallying, it now seems strange that only Audi looked to all-wheel-drive when FISA legalized it in 1979. After all, it was a lack of traction that had kept outputs (and therefore speeds) down.

But all-wheel-drive automobiles *weren't* an obvious answer—not by a long shot. In fact, Audi's plan was met with skepticism at first. The added weight of all-wheel-drive was seen as a major problem; Audi reasoned that added power, put to the ground with superior traction, would more than compensate. Reliability was also a bugbear, since rally cars have to be tough as nails and easy to fix;

The road version of the Audi Quattro Turbo Coupe set the automotive world on its ear. Turbocharged engine, luxurious interior and aggressive-but-understated styl-ing captured the eyes of those unfamiliar with the advantages of full-time all-wheel-drive. *Audi of America*

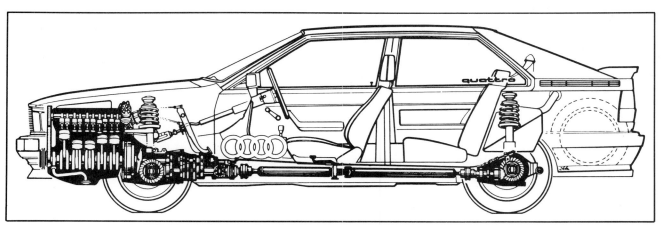

Inherited from the Audi 80 was an engine far ahead of the front axle. Despite this layout and strut suspension,

Quattros used all-wheel-drive to dominate World Rally Championship competition. *Audi of America*

57

John Buffum's Audi Sport Quattro continued the Buffum-Grimshaw victory streak beyond the mid-eighties. The short-wheelbase design had problems, but later development brought it back to front-runner status. *Audi of America*

Audi decided it would just have to develop reliable all-wheel-drive. And not the least of the worries was handling.

All-wheel-drive cars in Formula One, Indy and elsewhere had generated a few successes but a lot more gripes from drivers. Some complaints were well-founded, others were just a matter of unfamiliarity—the cars didn't set up into corners the same as rear-drive cars did, disconcerting many who had been raised in traditional racers. Perhaps most revealing, long after Audi made its all-wheel-drive aspirations clear, topnotch outfits like Lancia, Nissan and Opel continued developing new rear-wheel-drive racers.

The Audi Quattro Turbo Coupe was introduced to WRC racing in 1980 as a demonstration (course) car for the Algarve Rally. After the required copies had been built for Group 4 (the predecessor of Group B) competition in 1981, Audi struggled with the usual teething problems—most of them in the engine bay—and was never in serious contention for the season title. Quattros won three events outright, however, and led convincingly in so many others that by season's end, the Audis were dominant. Some drivers still complained that they were hard to drive at the limit, and they were often beaten on pavement—but pavement stages are not the rule in rallying.

Inside the Quattro Turbo Coupe

The Quattro Turbo Coupe was a souped-up luxury car—sort of a hot rod Lincoln in lederhosen. It was derived from Audi's relatively big front-drive 80 platform, and with that came an engine placed well ahead of the front wheels. The trouble was that the 80 was designed as a luxury autobahn tourer, not a backroad special, and its layout was less than ideal for what Audi had in mind.

The Quattro's driveline, too, wasn't perfect. The system was cribbed in part from the military Iltis vehicle that Audi made for Volkswagen, and it was virtually impossible to adapt sophisticated differentials—such as those FF was already building—to it. The original cars made do with a tiny center differential (often eliminated for competition) and open, manually lockable units front and rear. Torque was biased fifty percent front, fifty percent rear, even though it had been shown that most drivers preferred at least a small rear torque bias.

Yet the Quattro Coupe was tremendously successful. Why? Because it had all-wheel-drive, its herculean engine made the most of it; Audi's engineers were an on-the-ball bunch and the team had superb drivers who weren't afraid to learn something new. Audi simply bet that having all-wheel-drive, in conjunction with more than 300 bhp right from the start, would make up for any imperfections in the layout.

It was quite right. The World Rally Championship became an Audi showcase for the next three years. The team won seven of twelve events outright and took the manufacturer's title in 1982. It won five of twelve events and the driver's championship for Hannu Mikkola in 1983, and secured both the manufacturer's and driver's laurels in 1984.

In regional and club events the Audis fared even better, particularly with the top-class American entry of John Buffum in SCCA PRO Rally. Buffum, an experienced factory driver who'd already raced for Triumph, Audi and others, was an immediate success in his A1 (first-generation) Quattro.

"When we got the car, essentially in '82, it was clear we had a technological advantage that nobody else had," recalls Buffum—now out of driving and PRO Rally's series chief steward. "For '82 and I guess '83 we had a pretty easy run in the PRO Rally Championships. Nineteen eighty-four and '85 were a bit more difficult, 1987 was very easy, and 1986 was not too different." What Buffum doesn't mention is that he won the championship every year for Audi, and that only one man—Rod Millen in his one-off Mazda RX-7 4x4—ever came close to beating him.

Peugeot's 205 Turbo 16 4WS

By 1984, Audi had finally succeeded where so many others had tried and failed: the team demonstrated once and for all the value of all-wheel-drive. The Audi Quattro Turbo Coupe was the decisively dominant sort of racing car that all-wheel-drive had needed all along as a showpiece. Others had won races, but none had proved themselves so long nor so loudly nor so thoroughly. As to the matter at hand—WRC competition—1984 would also be the end of the Quattro's reign.

At the end of 1981, it became obvious to most of Audi's competitors that the era of two-wheel-drive was over. One of the first to accept the new situation and take fast steps to deal with it was Jean Todt, leader of France's Peugeot Talbot Sport.

Todt knew he could build a better all-wheel-drive racer than the production-based Quattro if he started with a clean sheet of paper. So even though his team won the 1981 series championship with a simple two-wheel-drive car, he announced that work was commencing on an all-new all-wheel-drive rally car for Peugeot. It was not until 1984 that his answer, the 205 Turbo 16, was ready to take on Audi.

Unlike the Audi, which had been developed from a mass-produced luxury car, the 205 Turbo 16 was designed as a race car first, a streetable homologation special second. (It was nevertheless a surprisingly pleasant machine in road-going form.) Though it shared the silhouette of Peugeot's 205 economy car, the 205 Turbo 16 featured all-wheel-drive, a mid-mounted turbo engine, an FF Developments center differential, unique tubular front and rear subframes, and many other features that would have been more likely inside a Porsche or a Ferrari than a French econobox.

May 1984 saw the introduction of the Peugeot 205 Turbo 16 at the Corsica Rally. Peugeots demonstrated superior performance from the start, but did not complete their first race. *Peugeot of America*

Homologation 205 Turbo 16s earned raves from European automotive writers and drivers. A boxy economy car silhouette kept the eyes of police elsewhere while a turbocharged, mid-engined four-cylinder propelled the car at tremendous speed. *Peugeot of America*

The first-generation 205 Turbo 16 lacked the tremendous horsepower and large aerodynamic aids of later cars. Performance proved excellent regardless. *Peugeot of America*

Austin-Rover and Williams Grand Prix Engineering teamed up to create the MG Metro 6R4. Ferguson Formula differentials and a double-overhead-cam V–6 weren't enough to make it a strong WRC contender. *Kermish-Geylin Public Relations*

The second-generation 205 Turbo 16's most visible changes were large front and rear aerodynamic aids. Weight distribution, rear frame structure, brakes and steering also got attention. *Peugeot of America.*

The result was immediately competitive, although it might not have been had Audi had better luck with its own new machine—a short-wheelbase, high-output version of the A2 Quattro called the Sport Quattro. The 205 Turbo 16 led but did not finish in its first two outings, then won the 1,000 Lakes Rally in Finland outright. Peugeot went on to dominate the rest of the year and the season that followed, gaining both the manufacturer's and driver's championships in 1985.

The Peugeot was effectively putting about 425 bhp to the ground by the middle of 1985, or nearly twice what had been usable in the two-wheel-drive cars of the recent past. Only Lancia's mid-engined, rear-drive 037 had been able to come close to the Audi and Peugeot, running with up to 450 (very

An evolution-model Peugeot digs in on the tarmac with Timo Salonen at the wheel. The codriver is either looking down at notes or can't bear to see what's coming up next. *Peugeot of America*

squirrelly) hp toward the end of its competition life. But by late 1985, even Ford and Lancia, at first rear-drive holdouts, joined the all-wheel-drive ranks in convincing fashion. Ford's promising RS200 was not yet homologated, but Lancia's new Delta S4 was up, running and winning.

The S4 was a car rather along the lines of Peugeot's 205 Turbo 16: a thoroughly advanced, mid-engined silhouette racer made to look like an econobox production model. With its supercharged *and* turbocharged engine, the new Lancia was poised to take charge in 1986, while the RS200, some months behind in development, looked to be faster still.

Nineteen eighty-six promised to be the most memorable rally season of the era, and tragically it was. The press had already taken to calling the WRC supercars Killer Bees, after the Group B class in which they ran. But in 1986, that name became terrifyingly accurate.

The Killer Bees

Few had stopped to question the idea of putting Indy car power inside machines that raced on mountain passes lined with spectators. There was even talk of a faster Group S designation for which just ten, not 200, cars would be needed for homologation. Formula One powerplants were often mentioned in connection with Group S, at a time when 800 bhp was not uncommon for a Formula One engine.

The warning came in the form of Ari Vantanen's serious crash in the 1985 Argentina event. But it was Portugal in 1986 that spelled the beginning of Group B's demise. A speeding rally car left the

In-flight characteristics are extremely important on rally cars fast enough to spend long periods in the air. This type of landing, and the rebound that follows, can be extremely dangerous. *Ford of Europe*

Rear torque bias let Stig Blomqvist kick the RS200's rear end out on slippery corners. Swedish fans bundle up in all kinds of weather to cheer rally stars on. *Ford of Europe*

The RS200 followed in the footsteps of other mid-engine, high-performance Fords like the GT40 and Pantera. The rally car left both of those behind in sophistication and drivability. *Ford of Europe*

The entire plastic rear body of the RS200 lifts to reveal a long-travel rear suspension and a turbocharged four-cylinder engine. From this angle, the Ford looks more like an IMSA GTO car than a rally entry. *Ford of Europe*

road on a fast stage and killed three spectators, a tragedy that surprisingly few had foreseen—though some teams had complained about the conditions at Portugal beforehand. All works entries were withdrawn after the accident.

But the season continued. Two events later, in the Tour de Corse, the popular Henri Toivonen crashed while leading the eighteenth stage in his Lancia. Toivonen's S4 burst into flames and both he and codriver Sergio Cresto were killed.

That was too much. FISA flurried to put the brakes on the WRC's rush toward speed. Group S was never mentioned again, and Group B was declared dead at the end of the year. For 1987, Group A would take its place, meaning that only cars made in lots of 5,000 or more could be eligible for the WRC championship.

Left in the wake of the Group B cancellation were a number of superb automobiles with little or no competition future. The legacy, however, was a permanent place for all-wheel-drive in the minds of enthusiast drivers.

Also left behind was a huge reservoir of 200-off all-wheel-drive supercars for lucky European hotshoes to choose from. Ford's RS200, Citroen's 1000 Pistes, Austin-Rover's MG 6R4, Peugeot's 205 Turbo 16, Audi's Sport Quattro and a few others were all available to drivers who could benefit from the brief, fascinating period of Group B racing. Unfortunately, few of these street-legal racing cars found their way to the United States.

SCCA PRO Rallying

Despite its huge popularity in Europe, rallying has always been more a club activity in the United

This is how an understeering Audi negotiates a dusty corner in Kenya. The front end is sliding and the driver has dialed in additional steering angle to get through the corner. *Audi of America*

Sharp stones can cut tire sidewalls if drivers aren't careful. It is hoped that the bull bar on the front of this car will prevent serious damage in front-end collisions. *Audi of America*

Chad DiMarco and his Subaru Sports RX three-door Coupe hustle through a corner during the 1988 SCCA PRO Rally Group A season. DiMarco's Subaru-backed effort brought the Japanese company back into the hunt in PRO Rally competition. *Subaru of America*

A forward-mounted engine and a fifty-fifty torque split give the Subaru RX–3 coupe considerable understeer, like that in the Audi Quattro and front-wheel-drive rally cars. *Subaru of America*

States. "Rallying has a couple of problems that don't fit into the U.S. way of doing things," says John Buffum. "One is the size of the country; the distance of the people from the good rally roads. The other thing is the vast number of small sports that are easy to put on TV because they only require one or two cameras: I'm talking about tractor pulling, beach volleyball, things like that. Even though it's exciting, it's costly to photograph a rally.

"Rallying and Formula One are the two biggest motorsports in Europe, and thus in the world. In the number of spectators [of WRC versus PRO Rally events] there's a tremendous difference, and of course, the better support levels there are, the better the competition and the better the athletes."

Which is not to say that SCCA PRO Rallying lacks anything in excitement or competition. But perhaps you'd rather find that out for yourself. Here's what you'll need to know.

Organization

There are two main grades of PRO Rallying, National and Divisional, with further classes and divisions within each. Basically, National races are for seasoned veterans, while Divisional events are geared to competitors who don't have the experience, time or money to compete at the higher level. National PRO Rallying is a series on par with IMSA GT, NASCAR Winston Cup or SCCA Trans-Am; Divisional PRO Rallying is more of a "feeder" series, like SCCA Formula Russell or IMSA Firehawk.

No more than a dozen National PRO Rallies are held in a season. An SCCA National PRO Rally must

The PRO Rally Subaru Coupe shows a fair amount of body roll when hustling through a tarmac section. Rally suspensions must be soft enough to soak up bumps on dirt stages. *Subaru of America*

Lots of forward dive under braking comes from soft springing. Chad DiMarco nevertheless covers ground quickly on this tarmac stage of the Olympus Rally. *Subaru of America*

be at least 250 miles long with 100 miles of special stages, but usually it winds up quite a bit longer.

Divisional PRO Rallies aren't quite so serious as Nationals, but they're not walks in the park, either. Even the simplest event, a Coefficient 1 competition, must have at least two special stages with a total distance of ten to twenty-five miles. (For the inexperienced rallyist, these might be the toughest twenty-five miles of his or her life.) The same rules for organizers of National events must be used as guidelines at the Divisional Coefficient 1 level, but some variations inevitably occur. The complete format and regulations of any Coefficient 1 rally will be made clear to the entrants in advance, and it's best for even the experienced rallyist to study them carefully.

Coefficient 2 rallies are a step up the ladder. Each has at least five special stages and a cumulative stage distance of between forty and seventy-five miles. Events at the Coefficient 2 level adhere much more stringently to the rules and regulations set forth for National rallies.

Coefficient 3 rallies are the longest and most serious in the Divisional series. Total stage distances are sixty-five to 100 miles and the adherence to competitor, organizer and format regulations is strict and exact. Coefficient 3 rallies can be run concurrently with, or score points toward, National events if the series chief steward, the PRO Rally board and manager, and the event's organizers wish.

Driver eligibility

To compete in a National PRO Rally, you must be at least eighteen years old and have the following:
• A current SCCA National PRO Rally license (or Fédération Internationale de l'Automobile driver's and entrant's licenses at FIA-sanctioned events).
• A valid, current automobile operator's license.
• A current physical examination as prescribed by the SCCA.
• Proof that you have completed at least one Divisional PRO Rally at the Coefficient 2 or 3 level within the last two years.
• Proof that you have successfully finished an SCCA PRO Rally school.
• An automobile conforming to the regulations for its class.
• Proof that you have fulfilled any special conditions as deemed necessary by the series chief steward.
• Suitable proof of insurance and registration.

Divisional entrants must be eighteen or older and have the following:
• A valid driver's license.
• An SCCA membership.
• A Divisional PRO Rally license or a combined Divisional-National license. (To get a Divisional license, applicants must attend an introductory seminar, which is held before each Divisional event, or attend an approved PRO Rally school.)
• A legal automobile.
• Suitable proof of insurance and registration.

The production version of the late-eighties Mazda 323 GTX appeared to be little more than a pleasant three-door economy car. A turbocharged twin-cam engine and full-time all-wheel-drive lurked under its unassuming skin. *Mazda of America*

New Zealander Rod Millen gets ready to drop a front wheel into a hole as his Group A Mazda 323 GTX heads for another PRO Rally win. *Mazda of America*

Vehicle eligibility

For National events, vehicles race inside strict classes. Divisionals usually run everyone in a single class, although the organizer can designate separate Production classes if desired.

Whatever the class and wherever the event, the SCCA demands some things for every vehicle:

• Regular safety and signaling equipment like headlights, taillights, stop lamps, rearview mirrors, horn, windshield wipers, turn signals, foot brakes and handbrakes must be in good working order.

• Tires must be in good condition on both sidewall and tread. Minimum tread depth is $\frac{2}{32}$ inch, measured across the complete tread surface. Protruding metal (as in studs) is prohibited in any form.

• All auxiliary driving lights must be wired to turn off when the regular headlights go to low beam. Furthermore, the base of any auxiliary light can't be more than an inch higher than the lowest point of the windshield.

• Flexible mud flaps must be fitted at all drive and rear wheels.

• A roll cage conforming to current SCCA or FIA specifications must be fitted. This is generally a mild steel or alloy steel fabrication, its thickness dependent on the car's weight.

• SCCA-approved five- or six-point occupant restraint systems, dependent on seating angle, must be provided for both occupants, and occupants must wear them at all times.

• Power door locks must be made inoperative. Power window units can be removed if desired.

Sunroofs, T-tops and moonroofs must be fixed in place. Glass and plastic roof panels must be replaced by metal of equal or greater thickness. The front windshield must be of laminated safety glass. Vehicles must be driven with windows raised, or fitted with approved safety netting.

• The vehicle must not produce noise in excess of eighty-six decibels, measured from fifty feet away, when run to 4000 rpm, held momentarily and returned to idle.

• Onboard and easily accessible must be an approved first aid kit (Johnson & Johnson Auto First Aid Kit or equivalent), three DOT reflective triangles and a halon or dry chemical fire extinguisher of 10 B:C rating or better.

• The hood must be secured by external pins; all other securing devices must be made inoperable.

Other regulations, depending on the vehicle, might include the battery mounting, fuel system shielding and so on. And whatever it says in the rulebook, grossly unsafe equipment can be ruled illegal on the spot. The organizers and promoters have the right (and responsibility) to demand any changes they feel are necessary in the name of safety.

In addition to following these regulations, vehicles must conform to the specific regulations of their individual classes. The vehicles eligible for National PRO Rallies are divided into five groups: Open, Production, Production GT, Rallytruck and Group A.

Open

This is the least restrictive class. Cars must be closed-bodied, four-wheeled, street-registered vehicles based on a production model recognized by the National Automobile Dealers Association (NADA) Official Used Car Guide. The engine must be based on a product of the vehicle's manufacturer. Brakes, suspensions, wheels and tires are free of restrictions; visible exterior body panels must be identical in appearance—but not composition—to the originals. Airfoils and fender flares can be added at will. All glass but the windshield can be replaced by Lexan of similar or greater thickness than original.

Production and Production GT

Eligible models must have been offered for sale in the United States at a rate of 1,000 units per year or higher. The SCCA defines a model as "a basic manufacturer's designation (e.g., Dodge Shelby Charger, Dodge Omni GLH, Volkswagen Golf GTI, Ford Mustang SVO, Chevrolet Cavalier Z24; a model is a specific vehicle and not a general category)." The model must be listed in the NADA Official Used Car Guide and must be of the current model year or six years previous.

Engines for the Production GT class must have an adjusted displacement of 2350 cc. Adjustments are figured as follows: for rotary engines, multiply original displacement by 1.8; for turbo- or supercharged engines, by 1.7; for all-wheel-drive vehicles, by 1.3; for three- and four-valve-per-cylinder engines, by 1.1; for pushrod engines, by 0.8. All other vehicles must run in Production class. Each vehicle must also have a full shop manual present at inspection.

No more than four additional headlights may be mounted, but type of glass, bulb and reflector are not specified. If bigger radiators are normally offered for the same model, one may be fitted. A single oil cooler may also be fitted. Elements that regulate the quantity of fuel to the engine are generally free of restrictions, but elements that regulate the quantity of air are not. (There's no boost limit on turbos or superchargers, though, and emissions equipment can be removed.) Exhaust systems behind the stock manifold are open to choice, but they must end behind the driver and free of the car's body. Any gearbox or final drive normally offered for the car is allowable. Shock absorbers, springs and MacPherson struts may be

With his fast 323 ride, Millen has become the new top man of SCCA PRO Rallying. All four wheels of the car are kicking up rocks as they scrabble for traction on a forest road. *Mazda of America*

A Japanese entry in the Olympus Rally heels over hard in a tight tarmac corner, fully extending the right-side suspension. Driver Okudaira's home-market sedan won't be this clean for long. *Subaru of America*

replaced by others of similar dimensions and identical mounting points. Some reinforcement of mounting points may be legal.

Any wheels and tires that mount directly to the stock hubs and fit inside the wheelwells are legal, provided they're a maximum of 6 inches wide or the original equipment width, whichever is greater. Brake linings are not controlled but swept area must remain unchanged. Clutch lining, pressure plate and flywheel are not restricted, but the flywheel must be of the same material as the original and fall within specified minimum weights. Unless offered as original equipment, limited-slip differentials are prohibited in Production class. Production GT cars can fit limited-slip or locked differentials as desired.

Seats, steering wheel and gauges are free of regulation. Carpets may be removed but quilting, dashboard panels and soundproofing material may not. Cruise control, sound system, antitheft devices and air conditioning devices may be removed; other options generally cannot. Passive restraint systems, including airbags, must be removed.

Rallytruck

The two-wheel-drive Rallytruck class generally specifies minimum weights (1.2 pounds per cubic centimeter for fuel-injected trucks and 1.1 pounds per cubic centimeter for carburetors), a maximum displacement of 2600 cc, a full roll cage, and modifications roughly like those allowed in the Production and Production GT classes. Engine balancing and open exhausts are allowed.

Group A

Group A rally cars must meet the specifications of the FIA Sporting Code for Group A automobiles, Articles 251, 252, 253 and 255 of Appendix J. The only exceptions are for fuel quality, which is not regulated; trunk lid fasteners, which are optional; and belt and roll cage specifications, which must conform to SCCA, not FIA, specifications.

What it takes to go PRO Rallying

It's not likely that the novice rallyist is going to want to take advantage of every modification

The ALCAN 5000 rally became a major media event in the mid-eighties, drawing factory-backed entries. All-wheel-drive and high ground clearance made up for a lack of overall horsepower in the Subaru wagon. *Subaru of America*

allowed. A rally car, like any racer, can become a money drain in a big hurry unless you make a realistic appraisal of what you'd like to have versus what your ability and experience merit. For one thing, your competitors at many Divisional events will be in the same boat you are; some won't have equipment much better than yours even if you've only met the legal requirements.

The absolute bare-bones requirements for a PRO Rally entry are a legal vehicle, proper entry forms and licenses, proof of registration and insurance, an approved vehicle logbook, an odometer capable of measuring to hundredths of a mile, and a driver and codriver who won't kill each other when everything hits the fan at once. (Contrary to what you might think, this last requirement can often be filled by husband-and-wife teams.)

Next-to-essential is a skid plate. There's no rule that says you have to have one, but the underside of your car—including the all-important oil sump—will take a heck of a beating if you don't. If you run without a skid plate, chances are you'll do more dollars worth of damage in your first rally than it would have cost to fit a good one in the beginning. Premade plates are available for a few vehicles, but mostly they're fabricated from aluminum by the car owner or a trusted shop.

More serious and seasoned competitors generally do everything the rulebook allows. As in any other racing series, when it comes time to start delivering the goods, you have to give attention to

every detail. At this point, however, the idea is to have a sponsor to help with the bills.

There are few hard-and-fast rules to setting up a rally car, though it's generally agreed that unlike road racing, rallying does not require lots of negative camber because of the relatively small side loads involved. Most teams stick to between one half degree and one full degree of negative camber up front, dialing in less for tires with extra-stiff sidewalls.

There's some debate on the best rear toe settings, but in general dead-on rear alignment will give stable straight-line handling and predictable cornering. Moving to a toe-out attitude makes the car respond faster to steering inputs, but also makes it twitchy during the transition from straight line to corner. Rear toe preferences thus vary greatly from driver to driver. Front toe is generally set slightly out.

Though your chances of success get better in rallying as your equipment improves, rallyists are quick to point out that a good driver in a lousy car can almost always beat a lousy driver in a good car. It's the driver that makes the difference.

Driving an all-wheel-drive rally car

One other big help is all-wheel-drive. Guy Light probably knows that as well as anyone: not only did he drive Volkswagen's Rally Golf with viscous all-wheel-drive in its debut at the 1989 Pikes Peak hillclimb, he's been rallying all-wheel-drives since

AMC's Eagle with viscous full-time all-wheel-drive wasn't generally regarded as a performance car, but

Guy Light had considerable success with one in PRO Rally regardless. *Light Performance Works*

Combine a tall Jeep CJ, a killer 400 inch Chevrolet small-block V-8 and a snowy PRO Rally stage, and you've got Guy Light enjoying himself immensely. *Light Performance Works*

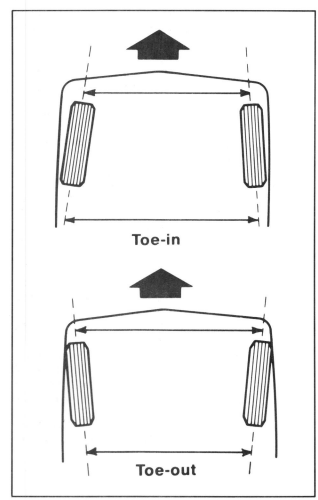

Toe-in

Toe-out

Toe-in and toe-out at either axle can dramatically affect handling behavior. Actual toe settings are fractions of a degree, not the four or five degrees shown in this drawing. *Toyota*

long before Audi made it popular to do so. Since Light was originally an off-road racer, it was only natural that he should show up for his first-ever rally in a four-wheel-drive International Harvester Scout. The Corvette engine inside was maybe a bit less natural.

The Scout eventually gave way to Saabs and then to a Jeep CJ-7 in 1977. That Jeep took an overall win in only its second outing, a twenty-stage event, and took Light to a third-place overall standing in the National series. Later fitted with a 400 ci Chevrolet small-block, it could do 140 mph, according to Light. There's got to be a record in there somewhere.

The most amazing thing about the Jeep, however, was its safe handling. "We did 22,000 race miles in it and never rolled it over," says Light. "We were never even in the woods with it!" Though Light is well known as one of rallying's best drivers, there's no doubt that all-wheel-drive also had something to do with keeping the tall, short-wheelbase CJ-7 right-side up.

Light's experience with the new Volkswagen just reminded him of what he'd learned in sport-utilities. "With four-wheel-drive you can go into a corner, and if you get in too far or if the car starts to feel like it won't hold traction, you can literally just accelerate on it. Point the car where you want it to go and it'll go there. That'll work until you get to a point that's so unbelievably slippery—like ice or something—that it wouldn't hold.

"Inherently, if the car starts to get out of shape, you can pause for a second, see how far out of shape it's going to get, and then accelerate on that issue pointed in the correct direction. It'll *go* that place. Which is very nice—it says you can go into a corner maybe deeper than you should, or into an unknown corner such as in rallying, and get yourself through an unusual situation."

Speaking in more general terms, Light also mentions that "a four-wheel-drive car cannot be driven showy—a four-wheel-drive car driven showy is slow. A four-wheel-drive car driven on the absolute, immaculate line without ever being showy is *unbelievably fast,* and very safe. You can't get them all sideways all the time and expect them to be fast. It tends to lend itself toward my style—I try to be real precise, executing each corner with the apex and the line and the whole bit exactly like it's supposed to be. Pikes Peak is particularly demanding in that respect, because you start at 9,000 feet—which means your motor's already a wimp—and you're going to above 14,000. To get a good time at Pikes Peak, especially with a normally aspirated car, you have to be extremely precise . . . take the shortest possible distance with the highest possible maintenance of speed. That's what a four-wheel-drive car does so well—it maintains your average speed."

Or take the observations of Robby Unser, the latest driving sensation from New Mexico's famous Unser clan. In winning the 1989 running of Pikes Peak with an all-wheel-drive Peugeot, Unser came to have a lot of respect for what four powered wheels can do: "With two-wheel-drive cars you've got to brake, get in, and use the back end to steer the car—you place it on the road and use your throttle. Your braking with an all-wheel-drive is much deeper so you drive right in. Also, on an open-wheel [two-wheel-drive] car you have a brake bias you've got to set all the time, while the all-wheel-drive stops at every corner the same no matter what. With all the wheels spinning it just pulls you right through. Where you're sideways into a corner in two-wheel-drive, all-wheel-drive just pulls you to the other side of the corner."

And finally John Buffum, citing experience with his Quattros, agrees with Light and Unser: "In a rear-wheel-drive car—in fact a rear-wheel- or a front-wheel-drive car—you tend to throw it into the corner more. In the front-wheel-drive car you charge into the corner, turn in, and use some left-foot braking to kick the tail out while you power on and power out of the corner sliding sideways. In the rear-wheel-drive car you might pendulum the car in, or in any case turn in and get on the power. That would kick the ass end out and then you could power your way around the corner spinning the rear wheels. With the four-wheel-drive you treat it a bit more like a racing corner." That means smooth, slick and with as little drift as possible. Buffum continues: "Again you may use left-foot braking, especially if you have a turbocharged car to keep the turbo boost up, but left-foot braking is not absolutely essential."

Buffum also notes that it took a couple of rallies for him to figure out the right style for all-wheel-drive, at least the sort the Quattro offers. "It does understeer—the Audi always understeered. Well, you just have to have the attitude of the car better when you come starting through the corner. You need to know you're in an understeering type of car, and with some left-foot braking you can get the handling reasonable."

Left-foot braking isn't a mandatory skill even for front-drive cars, where it seems to do the most good, but it's a good thing to know about regardless. The idea is as simple as it sounds: While leaving your right foot mashed on the gas, you squeeze on the brakes with your left foot.

Left-foot braking does a number of things: It creates a pseudo-limited-slip action by forcing all the drive wheels to pull against a similar strain, keeps the engine (particularly a turbo engine) up in its peak power range, and cuts out the time wasted shifting a foot from gas to brake and back. Most important, though, it modulates the car's balance. When one applies more brake, more weight is transferred forward, the front slips less and the unloaded rear slips more. In other words, more brake means more oversteer. There's danger, however, in overdoing it: too much braking can saturate (overload) the front tires, causing instant understeer just when you don't need it.

There's also one more attribute of left-foot braking that drivers like. "A left-foot braking application has a tendency to make the car more poised, and it doesn't lean as much," says Guy Light. "The word that most people use is that it *settles* the car." In fact, the only real drawback to left-foot braking, at least from the rallyist's point of view, is that it leaves you fifty percent shy in the leg department when it comes time to shift gears. The answer is matched-rev downshifts, which can be tough on the gearbox. Fortunately, the relatively low outputs involved in Production and Production GT rallying prevent much damage.

Rallying can be as expensive as you want it to be. For example, Buffum estimates that his first two Quattros ran about $125,000 each. Even in Production racing the costs can add up quickly: disassembly and inspection of parts, rerouting of brake and fuel lines, rewiring of electrics, fabrication of pans and cages, replacement of rotating items, a touch of engine balancing and so on all add up fast.

Many top teams regularly replace things like axle shafts, U-joints, gearboxes and engines, although the less wealthy have proven that that's not always necessary. Guy Light, in particular, has noticed that even things like a clutch can last a whole season if driven right. With his factory Volkswagens, he's found that very little major maintenance is needed. "The serious parts of a VW stay together very, very well . . . the actual engine and driveline. But those odd things that if you were a pavement racer you'd never have go wrong are the kind of things we beat to death because of the tremendous pounding." He mentions switches, fasteners and suspension bits as the things to keep an eye on.

What that means is that rallying can also be relatively affordable. When you are forced to put out a huge budget to even be competitive in many racing series, you can bring in a Production car in Divisional or even National rallying for much less than the top teams are spending. And since the series allows cars that are up to seven years old, you could even buy a used machine to start out with. A 1987 Mazda 323 GTX might be coaxed into just as competitive a platform as a brand-new model—providing you're a good enough driver.

Codriving a rally car

There's an alternative to shelling out for your own ride right at the beginning, and that's codriving. How you find the driver is your business, but once you've gotten into the right-hand seat you'll probably get the best rallying education there is. A good many top drivers started there, quietly learning about how a car should feel, where it should be and what the next turn would probably look like before they attempted to drive a car themselves.

One codriver who doesn't want to move on to the driver's seat is journalist Jean Lindamood of *Automobile* magazine. She doesn't think she can be the best driver in the world, so she's shooting for being the best codriver instead.

"I do it for the thrill of it and the speed," Lindamood says. "Co-driving is the only way you can ride in a race car and not be at the wheel but be as important as the driver.

"The job of the co-driver is to not get lost, and to check in on time. And I've raced enough myself to see when even a good driver is out of control. So part of the co-driver's job is also to tell when your driver is a little bit out of control and bring him back."

Bring back an out-of-control rally driver? This is supposed to be fun?

Lindamood continues: "You have to pick your drivers very carefully. I was determined to be as good as I could be so I could get the best drivers. I'm not interested in having someone kill me. I haven't been in the kind of crashes that say Tom Grimshaw has been in with Buffum. Furious crashes . . . you know where every body panel of the car is destroyed, and he's up in a tree. *Cartoon* crashes. I've

been in two major crashes. I can't even call them major. In one I went off the edge of the earth, just went over a cliff. But it happened to be a *short* cliff, and we went down the bank and ended in the river on our nose, with the engine running, hanging from our belts. And the other one [had us] up in a tree hanging over a river, in a truck. That one hit so hard the rollcage ripped a hole in the floor right behind my seat, but I was just sore. I've never been in any endos—knock on wood—and no rollovers. I really just take rides with drivers I trust.

"I've never raced with anyone who'd never rallied before. But if the guy's a Group A champion, you know that he had to do something to get there—you know he had to bring his car through. I rallied with Stirling Moss; I rallied with Malcolm Smith. These are guys who didn't get old being stupid."

Where did Lindamood learn enough to deserve her first ride? "I went to a co-driving school put on by the SCCA, and Ginny Reese was leading it then. The first co-driving I ever did was as a journalist in a press car driven by Rod Millen's brother, Steve, who's a real hot driver, too. And I was determined to do it well. So I went to Ginny and she gave me about ten hours of instruction in her kitchen.

"I spent a lot of time one-on-one just getting it all hammered into me. There's a lot of subtlety to it, a lot of detail. I decided that I would learn how to do it well enough that people would want me to be their co-driver based on merit alone. So that's how I

Walter Rohrl was on his way to setting a new overall record at Pikes Peak in 1987. On the other side of the gravel-strewn corner was a sheer drop that could quickly ruin Rohrl's morning. *Audi of America*

approached it, and I had a knack for it. And I got good enough that I got regular, good rides.

"I try and tell the driver at the beginning that I'm going to make mistakes at some time during the course of the rally. So I'll try to do it right away, and my goal is never to make one big enough that they'll talk about it in the bar later."

Pikes Peak

Zebulon Montgomery Pike never climbed Pikes Peak, but he was the first white American to see it, and his name stuck. The mountain looms tall at the top of Colorado, and maybe even a little taller in automotive legend.

The first car up the hill made it in 1902, in equal parts driven, pushed and pulled by two intrepid Denverites named Felker and Yont. All three (Felker, Yont and the Locomobile car) nearly died in the day-long process, but in the end they made the grade.

The trip got a lot easier in 1915, when local businessman Spencer Penrose opened the nation's highest toll road right up the side of the mountain. Even in those days, plenty of car-happy people were willing to spend the day chugging up the side of Pikes Peak—plenty, but not enough for Penrose. So he hit on a brilliant promotional scheme.

Ari Vantanen's oversteering Peugeot 405 Turbo 16 with 4WS charged up the Hill in 1989 trying to beat their record set a year earlier. Vantanen couldn't do it, and the win went to young Robby Unser in a similar car. *Peugeot of America*

Robby Unser seems to enjoy his way to victory at Pikes Peak behind the wheel of the all-wheel-drive Peugeot 405 with four-wheel-steering. *Peugeot of America*

"Forget the sheer drop-offs," Penrose thought. "Forget the air at the top that's about a third as dense as it is at sea level. Forget all that stuff like pea-graveled corners, no banking, and no guard-rails anywhere. I'm having a race up Pikes Peak and I'll pay cash to the man who gets to the top fastest."

In 1916, that man was Rea Lentz in his Romano Demon. Twenty-two-year-old Lentz whipped the likes of Barney Oldfield and Hughie Hughes with his homebuilt special, took the cash and faded away into obscurity.

Later on, Pikes Peak became the second greatest American race, and the second oldest, too, outdone only by the Indy 500. Every year on the Fourth of July weekend, the most daring drivers and willing manufacturers gave the hill their best shot. They had one 12.42 mile chance at winning this race, and a single screw-up meant anything from last place and no bragging to a one-way ride to Colorado Springs, 14,000 feet straight down.

Pikes Peak was still a unique spectacle in 1989, and it was going stronger than ever. For one thing, it was one of the last places where the legendary WRC Group B cars could still race unfettered. In fact, the cars that ran in 1989 were third- and fourth-generation descendants of the true Group Bs. They were 700 hp monsters more like the aborted Group S cars would have been.

Rally cars went a long way toward restoring the Race to the Clouds to its former glory and beyond. For a while in the sixties and seventies, things weren't going so well. Coverage and attendance were slipping, in a process that started when the big domestic companies withdrew from the Hill and moved to other places. Through sheer loyalty and competition, the hard-core nucleus of Pikes Peak survived declining public interest until Predator Carburetors was lured into sponsoring the race and things began looking up—so to speak.

Outside Colorado, however, the hillclimb had died. What had once been ranked with Indy, Daytona and Le Mans was now little more than a local title fought over by fiercely dedicated but largely unappreciated drivers. Then came 1981.

That was the first year of the Rally Class. It wasn't popular with the established order on the Hill—the little rally cars meant that some of their own number had to be bumped from the grid—but it was just the medicine Pikes Peak needed.

In Europe, at least, the name Pikes Peak and the Race to the Clouds still carried a certain air of mysticism. And Audi, already engrossed in dominating the world's best dirt road rallies, saw the Hill as one more race that its new Quattro might win. The man at the wheel was Audi's secret agent, John Buffum. "I hesitate to say this, but it's true," he confesses. "Audi—or however you want to say it, we—turned Pikes Peak around. It had been in the doldrums, and we came with this small, street-type turbocharged four wheel drive and we blew up the

hill faster than all these stock cars. Everybody, well . . . they couldn't believe it. Then we came back and did practically the same thing the next year, and things started to mushroom."

Audi and Buffum did turn Pikes Peak around. Other Europeans soon came, bringing with them journalists, photographers and lots of free publicity—just what the race had needed.

The secret was Audi's all-wheel-drive. Other Hill racers—some of them Jeeps, some homebuilt race cars—had had all-wheel-drive before, but none made such good use of it as the Audi. The regular machines for the Hill were off-road-style rail buggies with Porsche power, thundering V-8 stock cars or purpose-built open-wheel rear-drive cars. All of them negotiated Pikes Peak sideways, with drama and flair and a lot of wasted speed. The Quattros just went up the hill. Fast.

First John Buffum, then Michele Mouton, Bobby Unser and Walter Rohrl took Pikes Peak for Audi from 1982 to 1987. In 1985, Mouton broke the open-wheel record and moved Audi's street-based racers into the first-place overall title, where the rally cars remained at the end of the decade. But Rohrl was not alone on the Hill. Peugeot was there with its snarling 205 Turbo 16s, just waiting for him to take the air route to Boulder. He didn't, and in 1987 the Audi sweep continued.

Peugeot came back in 1988 with the 405 Turbo 16 with 4WS, an evolution of the 205 that had been so successful in Group B's salad days. With all-wheel-drive, four-wheel-steering and gobs of force-fed horsepower, Ari Vantanen took to the Hill in 1988 and set a time of 10:47.22, nipping Rohrl's old record by just 0.63 seconds.

Any race with corners called Devil's Playground, Bottomless Pit and Boulder Park is bound to have some devoted followers. Perhaps the best known of these have been from the Unser family: Louis, Jerry, Bobby, Al and now Robby, the fastest of them all.

Louis and Jerry first made the Unser name known to race fans, and they did it on the Hill. But it was Robby, at the wheel of a Peugeot 405 Turbo 16 4WS, who won the race in 1989, just missing Vantanen's record time but, at least for a while, taking the perhaps even more important Unser family record. He did all that even though it was only his third race up Penrose's highway.

Like any Unser, Robby wasn't new to racing. He already had a strong position in the American Racing Series atmospheric Indy car show, and a CART ride was coming soon. But with all-wheel-drive or two-wheel-drive, Robby found that Pikes Peak has special thrills all its own:

"Pikes Peak has what we call hard spots—spots where it's worn down and you get really hard on the rocks . . . you can really lay down rubber on 'em. Anyway, if the guy who goes in before gets a wheel too far to the outside and throws down some

Mitsubishi's home-market VR-4 came on strong at the end of the 1989 WRC, including Kenjiro Shinozuka's seventh place finish in the Acropolis Rally. Jimmy

McRae took fourth in a similar machine, behind three Lancia Delta HF Integrale. *Mitsubishi of America*

The Galant VR-4 WRC racer added an engine good for about 300 bhp to the all-wheel-drive, ABS and semi-active suspension of the production car. New cars like

the VR-4 have put Mitsubishi into the same league as Honda and Toyota. *Mitsubishi of America*

Hustling through an ice-slick rally stage, Ari Vanta-nen's Mitsubishi-Ralliart Galant VR–4 seems relatively calm—there are no roostertails and a minimum of front drift keeps the car in line. *Mitsubishi of America*

dirt and loose stuff on the hard spot, well—let's just say that two wheel drive *or* all-wheel drive, you've got to watch for it!"

He also admits that despite the bravado of Pikes Peak racers, the danger of what they're doing is far from forgotten. "Every time you come close to going off you stop and think 'That wouldn't be *no* good at all!' But I don't think it'll ever happen, because I won't *let* it happen. It *can* happen, but you're going to have to prove it to me first."

No one's lining up to teach Robby that lesson.

Chapter 5

Road Racing

All-wheel-drive is nothing new to road racing; in fact, the world's first all-wheel-drive, the 1903 Spyker, was built for that purpose. But despite many false starts since then, only in the late eighties did the technology of all-wheel-drive really take hold in tarmac competition.

Showroom Stock

Showroom Stock racing came about for a number of reasons, including the need to provide an affordable series to competitors and the desire of manufacturers to see their production models competing head to head. In Showroom Stock, production cars race the way they're delivered off the showroom floor. In theory, at least, these cars are

little different from the ones you'd drive away from the local dealer after signing on the dotted line.

One way the SCCA ensures that its Showroom Stock cars are representative of normal production models is to specify a high homologation number. In this case, only cars made in projected annual lots of 5,000 or more are eligible. One-hundred-off race specials and high-priced, low-volume exotics need not apply. Specialty manufacturers—for example, the Saleen company and its hopped-up Mustangs—that can prove sales of 1,000 units per year and a reasonable dealer network can appeal for eligibility on a case-by-case basis.

Showroom Stock dates back to 1972, when the SCCA started a series for absolutely stock vehicles

The Archer Brothers Talon at speed. This unproven race car set a pole position the first time it appeared on a track. *Archer Brothers*

The roll cage is pressed right up against the roof of the Archers' Jeep-Eagle Talon TSi. The rules say you can't weld the cage in, but you can sure pack it in tightly enough to add stiffness to the car. Interior panels that interfere with the ideal position of the roll cage are surgically sliced to make room. The entire installation is clean and neat. *Archer Brothers*

with a selling price below $3,000. But the real turning point for the series was a twenty-four-hour race held at Ohio's Nelson Ledges on the first day of summer in 1980—the now-famous Longest Day of Nelson Ledges. The thought of a twenty-four-hour endurance race for basically stock automobiles captured the imagination of a number of magazines and, consequently, manufacturers. That first Longest Day—basically an amateur event—got nearly as much coverage in 1980 as the 24 Hours of Le Mans it was patterned after.

The first Nelson Ledges and another twenty-four-hour race at Mid-Ohio in 1984 proved there was a great deal of interest out there in Showroom Stock, provided the endurance angle was thrown in to keep things interesting. Most people figured they knew which car could beat which in a sprint, but seeing how long the cars could keep it up added new excitement. (Not all spectators agreed: "It's like standing by an on-ramp for 24 hours," said one in a contemporary car magazine.)

The Saleen Mustang in the background has a five-liter V-8; the Talon TSi has a two-liter four-cylinder. A stock Saleen has only about 25 bhp on the Talon, though, and the Archers usually kept the smaller car out front. *Archer Brothers*

Subaru used an all-wheel-drive Legacy sedan to break the world land speed endurance record previously held by Saab. The Legacy logged over 62,000 miles at an average of 138.78 mph. *Subaru of America*

The SCCA decided to sanction a complete Showroom Stock manufacturer's endurance championship in 1984, and its first sponsor was *Playboy* magazine. The Playboy Cup ran for a year, then *Playboy* pulled out and Cincinnati Microwave took over. (Cincinnati Microwave makes the successful Escort and Passport radar detectors.) The Escort Endurance Series was born.

In the first two years of the Escort series, Chevrolet Corvettes were the overall winners of every race. But for 1988, Corvettes were banned to their own series, the SCCA Corvette Challenge, and other cars came in to fill their place.

Depending on their potential speed, Escort Endurance cars are divided into three groups: GT for the biggest engines, A for the slightly less powerful and B for the rest. All Escort Endurance races are between three and twenty-four hours long, so pace and reliability become important in even the shortest events.

As mentioned, Showroom Stock cars are only marginally different from the ones available at any dealer in America—in principle. As we'll see, there's some difference between principle and practice, but the cars remain by and large true to that ideal.

The rules say that engines can be balanced, and shocks and coil springs are given some leeway as to composition but not mounting points. Extensive safety and fire-protection equipment is mandated, and bolt-on aftermarket wheels of original equipment diameter are allowed. Additional tweaks like crankcase and differential breathers are allowed, the car's ride height can be changed, air conditioners can be removed and brake pad material can be changed. Other than those things and a few more—Walker Dynomax mufflers can be replaced for the stock units, for example—the cars are indeed of the same specs as the one your dealer would be more than happy to sell you.

What that means is that perfect preparation and setup are the keys to winning in Showroom Stock. The top teams all start with factory-built machines, but they strip them down to the bare chassis and build them up all over again before one ever turns a lap. Each component is checked for perfect true and flawless composition. Each fastener is given extra care and a dab of threadlock. Every bushing and gasket is trimmed to absolutely perfect size. And any part not quite up to snuff is chucked out and replaced with one that can do the job.

Bobby Archer, half of the famous Archer Brothers driving team that dominated Escort Group A racing in 1989, gave a quick description of

Added safety inside the Talon TSi comes from window netting on the driver's side and a padded front roll cage bar. Rules allow the steering wheel to be replaced, but this one's right from the factory. *Archer Brothers*

A stock front passenger's seat is mandated by the rule-book. A racing seat, left, is a lightweight padded shell with holes built in for a multipoint safety harness. *Archer Brothers*

the process: "We strip the car down to nothing, and every part really has to get some attention. Number one is bringing the car up to rules. For example, the safety equipment—the location of the rollcage, making sure you can get in and out of the seat quickly, making sure it doesn't hamper your vision, and making sure that in case the car *is* in an accident that the rollcage protects things like the gas tank."

Tommy Archer, the other half of the team, adds, "We put a roll cage in and people go 'Man, you fit that in so you can't even tell it's there.' There's a rule that says you can't weld it to the car, but it doesn't say you can't press it against the car. So it's pressed against the car to where it can be felt." By that he means that the roll cage adds stiffness to the car's overall structure. That's a perfect exam-

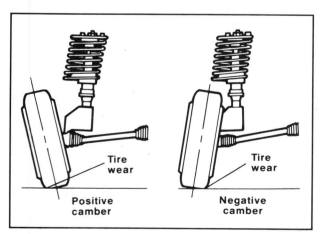

Camber is a measurement of the tire's angle to the road. Negative camber on outside tires aids in cornering. With negative camber, the top of the tire is angled toward the car.

ple of professional preparation that's within both the letter and the spirit of the rulebook, yet results in a car that performs noticeably better than one fresh off the dealer's floor.

Even playing with the composition of shock absorbers can have tremendous performance benefits, and the top teams go to great lengths to get them set up just right. "We had a pretty good idea of shock rates," Bobby mentions. "Bilstein obviously never [built shocks for a car like the Talon], and we had to take the shocks that were on the car, disassemble them, and send the shock bodies out to California to have Bilstein come up with a setting. Looking back, we've had about five progressions since then."

"We have shock valving anywhere from 280 rebound, 100 bump all the way up to 450 rebound, 145 bump," adds Tommy. "That's actually a fairly small range compared to what we do on the [Jeep Racetruck vehicles], but in endurance racing you don't have to be quite as particular because you normally don't go to your full capabilities."

All-wheel-drive comes to Escort Endurance

When the SCCA set out its predictions for 1989, it figured the Group A category would be a slugfest between Pepe Pombo's Nissan 300ZX, the Mazda RX-7 GTU and the new, unproven Shelby CSX. The Eagle Talon and Minnesota's Archer Brothers were nowhere to be found. Strange, perhaps, since Tommy and Bobby Archer shared the honor of having the most overall Escort wins going into 1989, with ten each.

What those with the SCCA didn't foresee was the Archers' last-minute bid for the Group A title with a brace of all-wheel-drive Eagle Talon TSi racers. And even if they'd seen it coming, the thought of an unproven all-wheel-drive sports car blowing

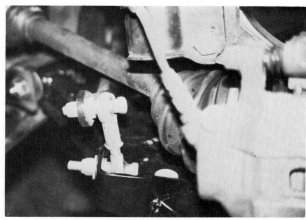

Normally, the strut front suspension of the Talon TSi can't be adjusted for camber. Tommy Archer added an eccentric bolt, usually used by frame shops, to allow adjustability. *Archer Brothers*

Even the wheelwells of the Archer Brothers Talon are clean and sanitary. The large link to the left of the photo is a cross-member of the independent rear suspension. *Archer Brothers*

the reigning V-6 and rotary coupes off the track probably wouldn't have entered their minds.

The Nissan-Mazda battle took place all right, but it wound up being a battle for second place that was distinctly overshadowed by the Archers' little Talons. In rapid succession came a Group A Eagle win at the 1989 running of Nelson Ledges, then a one-two finish at Mid-Ohio, a first at Lime Rock and another one-two at Mosport Park in Ontario. By the Mosport Park race in August, the Archers and their team of Talons were leading the series and never looking back.

One amazing thing about the Archers' success with the Talon was how quickly they were able to put it all together. They were building the race car while production was just getting started at the Chrysler-Mitsubishi Diamond Star factory, and the plant's start-up schedule didn't leave them much time to develop their racer. "Not so much as a cam cover gasket was available at first," says Bobby. But because of their experience in setting up other winning cars, they were able to turn a very early production Talon into a pole-winning race car in a matter of weeks.

The Archers are quick to credit their success to the toughness of their cars, their own experience as drivers and builders, and the great handling that the cars had right from the factory. But they also don't forget to mention the Talon's sophisticated viscous-controlled all-wheel-drive system. Bobby says, "We have to compare this car with the other cars that we're racing against, which are running from Honda CRXs to Camaros. Our braking is better—and I expect it to get better than it already is. We can out-brake just about anybody."

"I think if he thought back he'd probably say 'But then again, I've *always* out-braked everybody,'" quips Tommy.

"And the entry into a corner is quicker than I believe everybody that we're racing against," Bobby continues. "And where there's a series of tight S corners, a series of rights and lefts, we're quicker in there. I assume it's because of the AWD. The car responds quicker. There's corner 8-9 at Mosport, and we can literally drive through that where other people were either on or off the gas. You can really catch people there.

"Tommy and I started racing low-powered cars, and we've always had to be more aggressive in the corners. It [the all-wheel-drive Talon] allows us to be even more aggressive, especially in an endurance race where there's hundreds of pounds of rubber deposited on the outside of every racing line. With the Talon, that's no longer off limits. You just drive the car into the off-limits zone and drive through it with all the confidence in the world. The drawback to that is with us running the S-compound tire, the tire is fairly soft, and we pick up all that rubber. You end up with a severe vibration for half a lap while it all gets thrown back off."

"I think that the car," adds Tommy, "because of the capabilities of holding the road, tends to let you go farther into the corner even without you knowing that you're doing it. That was the one thing I noticed right away. Also, like at Lime Rock, a front-wheel-drive car would go up the hill and probably change lanes without turning the steering wheel, because you're going through a bump-steer mode. The car goes all the way to the top of the suspension and back down. With this car, you just keep your foot to the floor."

The added stability of all-wheel-drive also has a payoff in safety, which means life and limb, and the side benefit of keeping the car in the running when otherwise it might be out for good. Bobby uses one of their hired drivers as an example: "At

Nelson Ledges, in the rain, we were using basically slick tires, and one of the guys slid off into the grass. Usually in a race car if you drive off into the wet grass . . . well, you pick up speed! But he pulled it right out."

Tommy got a little closer to the action. Talking about his own experience, he says, "You slide off, even in wet weather, and you can accelerate back to the track. Whereas in most cars you just try to *make* it back to the track, with this car you *accelerate back to the track.*"

Preparing an Escort Endurance racer

In addition to doing thorough prep work—including completely disassembling and testing the entire car—Showroom Stock racers need to take every advantage they can from the SCCA rulebook. "You got to do what the rulebook says," Bobby Archer believes. "It says you can use 16x7 inch wheels for the Talon, with so much offset, and that's what we have. It states the minimum and maximum dimensions for rollcage tubing, so we go to that. It says the maximum amount of camber you can use, and we have that adjustment built in. It says you can use a Walker Dynomax muffler—we

used it. Everything that it says you can do we do. It says you can take the air conditioning out, we do it.

"And when you build your car to the rules, to the minimums and maximums, in theory you should have a competitive car. The Escort series allows for alternate brake-pad material—I think we're on like our sixth- or seventh-generation brake pad. Always look for something better."

Sometimes that extra edge isn't found just in the fastest pieces but in pieces that can be counted on to do the job. Tires are one example. Bobby says, "Goodyear may not be the most aggressive [tire manufacturer], but Goodyear has been honest with us. They provide engineers at the track who know what they're doing, and we are involved with their tire testing program. We've got a Goodyear test scheduled in a couple of weeks where we're going to try some new compounds *specifically* for the Eagle."

Since the Escort series stresses endurance, preparing for overall top speed isn't necessarily the best way to win races. Bobby explains that "the track surface, as far as adhesion, is the biggest thing that we have to work with; more than cornering ability. The car runs about an hour and a half on a tank of fuel. We want those tires to last about an

Despite minimal body roll, the inside front tire is lifting off the track during hard cornering. This would be more of a problem with front-wheel-drive than with all-wheel-drive. *Archer Brothers*

Tommy, left, and Bobby, right, discuss the effect of every change they make to the Talon, particularly if it concerns tires. Constant documentation is one key to their speed. *Archer Brothers*

The Mitsubishi Eclipse GSX should prove just as successful as its Eagle Talon cousin in racing. Dave Wolin Racing Inc. sent this one into IMSA's International Sedan (LuK Clutch Challenge) series for 1990. *Mitsubishi of America*

hour and a half at endurance racing speed. We're not looking for the ultimate fast lap, we're looking for the most consistent lap times during that fuel load."

Suspension tuning is a major part of any racing program, including the Talon's. But the Talon program was different in that the Archers were starting from scratch. There was no prior experience to fall back on with an all-wheel-drive endurance racer. "We liked the challenge," Bobby says. "I mean, we've won races in Jeep pickup trucks...we've even won races in a Chevy Cavalier, for crying out loud. The whole team just got together, and really we knew what the basics were—we had to make the car bulletproof."

All-wheel-drive or not, the Archers knew that their top priority was to get the car reliable and simply into the running. Fine-tuning the setup and learning how to get the fastest lap from the driver's seat had to wait until more experience could be gained on the track. As it turned out, prior experience at Sears Point, the car's first outing, helped considerably in the initial setup. The car took the pole right off the trailer.

Because of Escort Endurance rules, springs, sway bars and almost all the rest of the Talon's suspension bits had to be left just as they came from the factory—except that everything was checked and trued to perfection. Any less-than-perfect parts were tossed out and replaced with new ones. That just left settings of toe and camber to be played with. "We don't necessarily set it up for a corner; we set it up for a track," says Tommy Archer.

Tommy again uses Lime Rock as an example: "We gave away the one left hand turn and took all the rights. At [Mosport Park] it was the same way, but we had one long downhill left hander, so we had to leave some left hand turn qualities in it— which is camber. Most of the time, if it's a right-hand track, you'll see that the left side of the car has got a lot more negative camber than the right."

Normally there's no adjustment for camber on the Talon's front struts, but Tommy has found a way around that. "We build our own adjuster, a little oblong eccentric bolt. That's a normal alignment piece that frame shops use when someone comes in with, say, an Omni/Horizon that's not normally adjustable. Instead of replacing a $130 spindle, they put this piece in for $15, weld a little tab on, and then they can adjust the suspension back true.

"This car basically asks you for the same type of suspension settings the Corvette does. Most of the time, in order for the back end to follow the front, it requires toe in at the rear and usually a little toe out at the front to make it handle properly."

The effect of the Talon's all-wheel-drive remains something of a gray area when it comes to fine-tuning, however. In the middle of the 1989 championship season, Tommy Archer remembers, "I put a limited-slip differential in the back to give it more traction at the back, and it pushed. Something that right now we'd like the car to have is less traction up front. Right now the tires up front are doing stopping, turning, and driving, and the tires get overworked because of that. You're putting them

Team Mitsubishi-Ralliart U.S.A. was understandably eager to replace their successful two-wheel-drive Eclipses with the stronger, all-wheel-drive Eclipse GSX. Experience with the two-wheel-drive car meant much of the program's development work had already been done. *Mitsubishi of America*

through abuses that people on the street will never see, and when you're asking a tire to do more, very seldom will you get more out of it [as far as] mileage."

One thing the Archers have always stressed over and over for any series is the need to follow simple, good racing practices. These include not only thorough preparation but complete documentation of every change in hardware and lap times. Bobby explains: "We've raced against some teams where they have two cars. One car's got a BBS wheel and one car's got a Ronal wheel on it. Well, there's a quarter inch difference in total offsets. How can they ever compare anything? You're looking for hundredths of a second and you know, they're not doing it. You've got to watch *everything*."

Watching everything includes making sure that potentially worn-out parts are replaced at regular intervals. On top of planning for frequently changed suspension bits like front hubs, control arms and so on, the Archers have worked out a schedule for the rest of the car as well.

But while the brand-new Diamond Star factory was having trouble just keeping its production line stocked with parts, things that normally would have been changed regularly had to be pressed into extended service. One engine logged over forty hours at racing speeds before it could receive any major attention. A single transfer case, transmission and differential were used for more than forty hours without care. "You could never do that in a Camaro," according to Tommy. "[The Camaro teams] changed a rear end in each car in a twenty-four hour race this weekend [Mosport Park]. We ran a three-hour, a twenty-four hour, a twelve-hour, and a three-hour on the same differential and transfer, and they were fine. We just took them out because we felt sorry for them."

While all-wheel-drive divides torque between four wheels instead of two, thereby reducing loads on the driveline components themselves, it also has a lot more components that can fail. So it's hard to say whether the Archers' good luck with parts is due to the car's own toughness, the brothers' smooth driving style or the damping qualities of the Talon's viscous center differential. The last possibility is worth further study. If viscous couplings prove to significantly reduce driveline failures under racing conditions, that may be just one more advantage all-wheel-drive has in road-racing applications.

The price of playing with the big boys

Even with a tough car, Showroom Stock racing isn't cheap—at least not if you want to win. While Bobby Archer stresses that you can't put a price on Showroom Stock cars in general—the setup and equipment vary too much from car to car and race to race—he throws out $50,000 as a nice, round figure for someone who wants to construct a car like his. That's assuming the constructor already has a shop and staff put together.

That means the entire car costs less than a mediocre CART engine, and less than you'd spend building a car for any major open-wheel series. But remember, he's talking about a Showroom Stock car that costs only about $16,000 in a real showroom.

Driving the Showroom Stock Talon

"I expected a bigger revelation than it really is from the driver's seat," Bobby Archer says of the overall handling of the Talon all-wheel-drive. "That could possibly change from car to car. On the racetrack, the [Talon] drives like a front-wheel-drive car—a *superior* front-wheel-drive car. You use the throttle a lot, very similar to a front-wheel-driver. For example, normally entering a right-handed corner you would try to do your braking before you got to the corner, then tend to get on the throttle and get the car set up right. You continue to depress the gas pedal, and if it doesn't want to turn, you can usually let off a little bit and induce it."

Of course, there are things the all-wheel-drive car can do that a front-drive car can't. Bobby continues, "Perfect case in point comparing it with a front-wheel-drive car: You go into a corner too deep and the inexperienced driver backs out of the throttle big time. Well, [with front-wheel-drive] the front end stops and the back end keeps going, and you spin around. This car does not have that tendency. You let out of the throttle and it tends to steer in the direction that it was going. We have not spun one out yet, and we've run a whole bunch of racetracks and, I think, about eight different drivers."

Tommy Archer says, "In a rear-wheel-drive car, when you take off and accelerate hard, the back end will slide. [With a] front-wheel-drive car, you normally get torque steer. With all-wheel-drive it basically makes it so you can drive the car with one hand. The car is simply smooth. My first impression of it was at Sears Point Raceway, which is one of the hardest tracks in the country, and I drove all the way from Turn 11 to Turn 4 with one hand on the wheel, the other hand talking on the radio. They have a tendency to understeer under full power [like a front-driver], and what we're finding is that if you put your foot all the way into it, the car will push real hard. If you let all the way off, it will push. But if you accelerate gently, the car will track all the way through the turn.

"That's not true in every turn, and each track is a little bit different. [At Mosport Park] there was a turn that, if you went into it with your foot all the way off, the back end wanted to be loose. And if you went a little farther and let off, the front wanted to be loose. So you really had to be careful."

As to how you approach a race overall, Tommy says, "At all these tracks I've driven front-engine, front-drive, front-engine, rear-drive, and now this car, and I think you basically have to attack the track the same. The one thing about this car is that it allows probably for a few more mistakes.

"At Lime Rock, I went around a kink and I slid off the track, and I said, 'Oh, Wow, we're going to lose it.' But I got my foot in it and pulled it right back on the track and we only lost a couple tenths of a second."

The brothers also discovered how well the Talon's all-wheel-drive compensated for surface imperfections and weather. "It didn't rain," Bobby says of Mosport Park, "so we were really disappointed!" It didn't matter. The Archer cars finished first and second, and an Archer-prepped Talon finished third.

Driving all-wheel-drive road racers in general

When discussing the handling characteristics of all-wheel-drive road racers, remember that there haven't been enough of them yet to get the rules down pat. In general, while a fifty-fifty front-to-rear torque split may be fine for a relatively low-powered car like a Showroom Stock Talon, with larger amounts of power on tap engineers like to bias more torque toward the rear of the car. Stirling Moss' Ferguson P99 Formula One car was a good example: its torque was split so that about one-third went to the front and two-thirds went to the rear. The idea was to emulate the throttle-steering characteristics of a rear-wheel-drive car. Audi does the same thing on its own all-wheel-drive road racers, although its split is less radical, with forty-four percent of the torque going forward and fifty-six percent heading to the rear.

With an even torque split, all else being equal the front end does most of the work during cornering, since it usually carries most of the weight entering a turn. (Deceleration shifts the weight forward, which is necessary for big contact patches on the steered wheels.) If the car is designed with more torque going to the rear, then even under severe deceleration the lightly loaded rear drive wheels will still be doing their share. In fact, it is hoped that they're capable of doing *more* than their share; that way, by proper application of the gas pedal, the driver can induce throttle steer at will.

For the Archers, who already had extensive experience racing front-wheel-drive cars, the front-drive nature of the Talon's fifty-fifty split caused no problems. And as experience is gained in building faster and more powerful front-wheel-drive racing cars and drivers learn to cope with them, we just might see stronger all-wheel-drive racers carrying an even torque split and behaving more like front-drive cars than rear-drive cars.

Audi's road racers

Audi's rally Quattros had given all-wheel-drive its first big foothold in the performance game, but the very nature of rallying left some people unconvinced that all-wheel-drive had a place on regular roads. Those people saw rally cars as high-performance Jeeps, and they already knew that off-road vehicles were supposed to have all-wheel-drive. (They forgot that until the Quattro had proved its effectiveness in rallying, people thought rallying was more like pavement racing than off-

This phantom view of the Audi 200 Quattro Trans Am racer shows the roll cage and subframe tubes used to brace up the basic unibody structure. The fuel cell and dry-sump oil reservoir rest in the tail section. *Audi of America*

At Talladega Speedway on March 24, 1986, Bobby Unser set a new closed-course speed record for all-wheel-drive vehicles. The Audi's one-lap average was more than 206 mph. *Audi of America*

The speed-record Quattro 5000CS Turbo used a 650 hp intercooled, turbocharged, five-cylinder engine to move above 200 mph. The good aerodynamics of its standard body also helped considerably. *Audi of America*

roading and therefore *not* a good place for all-wheel-drive.)

In the middle of 1987, to end the confusion once and for all, Audi decided to take the Quattro concept into road racing. "We had already waited too long," says Dieter Basche, one of Audi's top engineers, "to show that even on a good surface, on a normal tarmac road, four-wheel-drive has its advantages. That was the idea to prove: that it's not only good on gravel and snow and ice but on a normal road."

It was also important that something be done to improve Audi's image in the huge American market, where the undeserved stigma of "unintended acceleration" still seriously dogged Audi's sales. (The U.S. government determined the cause of unintended acceleration in 1989: The owners who experienced it were standing on the gas pedal when they thought they were on the brake!)

Toward the end of October 1987, word was finally passed down that Audi would indeed construct an all-wheel-drive road racer. The SCCA Trans Am, a long-established sports car series, was chosen for its first assault, and the first race of the season would be at Long Beach, California, on April 16, 1988. That meant there was just half a year to get a team and car together for the coming season.

So while Trans Am had become little more than a silhouette series—meaning the cars merely had to *appear* as though they were based on production models—Audi chose to build its own race car from a stock platform. That platform was the Audi 200 (then known as the 5000 in America), a large four-door sedan quite unlike the sporty Merkurs, Mustangs, Z cars and Camaros that made up most of the Trans Am field.

Fortunately, Audi had already done a lot of the development work that could be carried over to the Trans Am effort. That experience came from the extensive rally program and in building an Audi 200 speed-record car for Bobby Unser to run at Talladega, Alabama.

Unser's black Audi sported a 650 hp, five-cylinder turbo and set a closed-course lap record for all-wheel-drive vehicles, averaging 206.8 mph. Audi used this car as the jumping-off point for its Trans Am effort.

Audi also decided to hire Bob Tullius' Group 44 team to run the cars in America, drawing on its vast knowledge of American road racing, established organization and veteran staff. (Group 44 is so well organized that its mechanics sometimes lounge around the pits in clean new overalls while everyone else wrestles with last-minute details. It's supposed to drive the competition mad, and it does.)

Inside the Audi 200 Trans Am racer

Brian Berthold, a Group 44 engineer who's gone on to work hand in hand with his counterparts at Audi, describes the basics of the world's first truly successful all-wheel-drive road racer: "A lot of the components from the rally program carried over, but there were a lot of new solutions required to both fit the rules and the chosen basis—the Audi 200. It was an Audi 200 unibody, not a tube-frame car; it was more like a stock chassis that we put a roll cage in. It was further than that but, for example, the rear control arms were stock; they still had the rubberized coating that the factory put on them for corrosion protection."

As is standard practice, the roll cage was designed not only with driver safety in mind but also to stiffen up the unibody and achieve some of the rigidity inherent in a tube frame. Not only were tubes added to completely protect the driver, for example, but they were also run forward through the firewall to brace the shock towers holding the production-style MacPherson struts.

Audi's tight production schedule demanded that a tremendous number of production pieces be used inside the stock unibody, though the rules did not. The gearbox casing, rear differentials, front and rear glass, much of the rear suspension and many other parts were straight off the production

Close, poorly paved street circuits were the 200 Quattro's specialty. Low overall speeds and poor surface conditions gave the Audis an extra edge. Trans Am Quattros were longer and wider than the competition, none of which were four-doors. Careful attention to aerodynamics helped overcome the large frontal area. *Audi of America*

89

To the surprise and dismay of the competition, four-door 200 Quattro Turbos were immediately competitive in Trans Am. As the season progressed, weight and other penalties were given to the Audis in an attempt to slow them down. *Audi of America*

line. "The original prototype for the Trans Am car still had opening doors and windows," Berthold recalls. "Naturally we did away with those for the actual race car, but [the race cars] were that close [to stock]."

To save even more time, many of the specially fabricated parts on the car were merely beefed-up versions of production pieces. While the car's Mac-Pherson front suspension was unusual in a series that allows Formula car suspensions, the struts were used because they were proven pieces that could easily be adapted to the production unibody. Wishbone suspensions would have meant building in new pickup points to carry them on.

Fortunately, there was always the experience of the rally cars to fall back on. A surprising amount of what was learned on the snow, gravel and pavement of the World Rally Championship was directly applicable to Audi's pavement-bound road racer. The engine, driveline and brakes, for example, were all basic adaptations from the Sport Quattro program.

Dieter Basche comments: "[As for] the drivetrain, we had a lot of that experience out of the rally car. You know a lot of the world championship rallies have a lot of tarmac, and even though the roads are rougher and the tires are smaller, we could use a lot of experience on the road racing car. What we had to adapt were all the suspension parts and the kinematics. The car had to be a lot lower using a lot bigger tires, and with the bigger tires you had much higher cornering forces and brake forces. That's what we had to learn."

Still, despite Audi's expertise in the formerly black art of all-wheel-drive and the availability of many already proven pieces, building the Trans Am Quattro was hardly a matter of slapping bits together and going racing. Some serious problems needed to be dealt with in the short time available, and many of them could have sunk the team before it started.

Just as with the Quattro rally cars, the road racer's engine placement was perhaps the biggest difficulty of all. "In Trans Am racing you're allowed to push the engine way back," engineer Berthold explains. "But because of the fact that we used the stock engine and gearbox, and you need the axles to come out along the front wheel centerline, we couldn't do that. We had to run it in the stock location, which was a disadvantage because then the engine is hanging out ahead of the front wheels."

With that determined, compensating for the engine location became the main thrust of the car's development. Berthold continues: "Everything you could do to make it light in the front end was to your advantage. Naturally we had to run some ballast, and as the season progressed they added

Smooth, fast road courses were supposed to be tough on the 200 Quattros, where traction was not as important and large frontal area limited top speeds. But victory came on these tracks as often as on road courses. *Audi of America*

weight penalties that helped the bias problem; naturally we could add that wherever we wanted to. But the main thing was the lightness of the front subframe [the unibody was replaced with an airy subframe ahead of the front wheels] and everything forward of the front wheels. Extreme attention to detail was paid, as well as locating things like the oil cooler and battery and so on in the rear of the car."

Audi's official press releases claimed that front-to-rear weight distribution was a perfect fifty-fifty for the Trans Am car. Loaded for racing, though, the cars were probably running closer to a still-respectable fifty-five percent front, forty-five percent rear weight ratio. But just getting even weight distribution is not enough in a racing car.

An even more important consideration is the car's polar moment of inertia. Recall that a car's polar moment of inertia is a measure of how tightly its mass is packaged toward the center of the chassis. In general, the closer the mass is toward the center of the car, the faster and easier the car will change direction. As they did with weight distribution, Audi and Group 44 found a satisfactory compromise here. Berthold says: "There's a tradeoff

there. What's more important? The weight bias or the polar moment? When we went to the rear of the car, we tried to make the polar moment not a complete barbell configuration. In other words, when we moved things to the rear we tried to keep them at or inside the wheelbase. But for the fuel cell, you really can't do that and we wouldn't. It was just one of the problems that you have to work around with the layout, because the only solution was to relocate the engine, which was not possible." In the end, through painstaking detail work, an excellent compromise was reached between polar moment and weight distribution.

The engine-forward layout also made for less-than-ideal weight transfer under cornering, acceleration and braking. Helped in part by the ready traction offered by all-wheel-drive, longitudinal (fore-and-aft) pitching wasn't as big a problem as lateral (side-to-side) motion. According to engineer Berthold, "The whole game [in selecting springing] is weight transfer, and again there was a paradox there. At the front you want to try to limit the weight transfer from side to side because there's so much of it up there. You do that by running softer [spring rates at the front] com-

On the way to becoming the Trans Am Driving Champion in 1988, Hurley Haywood racked up Audi's first win on the tight, hot Dallas road course. It was Audi's second race. Haywood ran smooth, restrained races to capture the title. *Audi of America*

pared to the rear. But because the engine's hung out like that, you can't go *too* soft, or under braking you'll have a big pitching moment." Between extremes of forward pitch and lateral loading a solution was found. "Once you go beyond a certain amount in the rear," says Berthold, "you'll just end up losing traction in the rear *without* gaining in the front, so the final configuration was just the result of testing and learning at the track and development."

Fortunately, the dynamics of running gobs of power through all four wheels had been thoroughly explored with the Audi Quattro Turbo Coupes and Audi Sport Quattros before the Trans Am project began. The engineers were dealing with more than 500 hp right from the start, and getting the right characteristics out of the driveline was essential if the driver was expected to make the best use of all that power.

The choice was made to go with a forty-four percent front, fifty-six percent rear torque split, and Berthold feels that was right on the money: "If

we gave it a little more rear, the cars had a tendency to be loose—to break the back tires too easily, because we were running the same size tires front and rear. If you go to fifty-fifty, the danger is just that you'll have too much power understeer, which in a race car is absolutely the wrong thing to have. If you're going to have a powerslide condition, you've got to have that slight rear bias to make it controllable to the driver, so he can steer it with the throttle.

"The fronts naturally are still doing a tremendous amount of pulling, but if you had too much *front* bias you'd only be emulating a front-wheel-drive car. To get right ideally in between, *really* at the limit of control . . . well, I'm not sure if you can, even if [the track] conditions were consistent enough. So if the wheels are going to break loose under power, it should be under the driver's control to steer the car at the back as well as the front."

When the Audis arrived at Long Beach, there was snickering from some of the opposition. The team showed up with a pair of four-door racing cars powered by a forward-mounted, five-cylinder, two-valve turbo engine, and they sported the full rocker-to-roof height of the production Audi 200. The competition, meanwhile, was running purpose-built plastic bodies shaped vaguely like production coupes that were wickedly channeled (for less frontal area) over race-built tube frames. Their V–8 engines were rumored to be about 650 hp strong. Few people looked twice at the Audis out of anything more than curiosity. What were the Germans hoping to prove?

The Trans Am Season

Audi's Trans Am Quattros arrived at Long Beach ready for their first race, quickly but beautifully prepared. Hurley Haywood, an American road racer and Audi dealer, was chosen to drive one car, and two German pros, Hans-Joachim Stuck and Walter Rohrl, were picked to share chores in the other.

Haywood had Le Mans, Daytona and Sebring wins to his name, as well as rides in everything from Indy cars to Group C and GTP prototypes to Audi Escort Endurance racers. Rohrl gained extensive experience with all-wheel-drive while sitting at the top of the World Rally Championship heap through the eighties. He also shattered the Pikes Peak hill-climb record in 1987 with a specially prepared version of the Audi Sport Quattro. And Stuck, a Le Mans winner, former Formula One driver and former Porsche works pilot, was simply one of the world's best and most adaptable circuit drivers.

Right from the start, the Audis amazed the crowd with their abilities. They were competitive and reliable in their first race, things that no other tarmac-bound all-wheel-drive racer had been. At one point, Stuck was running second before a

German rally ace Walter Rohrl proved just as adept on pavement as on snow and ice. Rohrl was perhaps the hardest charger of Audi's three-man Trans Am effort. *Audi of America*

Former Formula One driver Hans-Joachim Stuck traded driving duties with Rohrl as both continued racing in Europe during the American Trans Am series. *Audi of America*

Technologically advanced GTO Audi Quattro 90s were given many more modifications than the Trans Am cars. Smaller and more aerodynamic than Quattro 200s, GTO cars also had stronger engines. *Audi of America*

backmarker put him out of the race. Turnabout became fair play at race's end, however: Hurley Haywood was running well, but seemingly out of the hunt, when a major snafu removed the front-runners and left him with a second-place finish behind Oldsmobile's Paul Gentilozzi. A second place in its first race, no matter how it was attained, was more than Audi could have hoped for.

People who pointed to the missing front-runners at Long Beach were silenced at Dallas, the next race. When Willy T. Ribbs crashed under pressure from Walter Rohrl, the Audi moved into first place and began walking away. Just five laps later, Rohrl, too, fought with a backmarker and lost, bashing up his Audi too badly to continue. Haywood, meanwhile, had been calmly cruising for championship points and staying out of trouble. He was there to pick up the pieces and went on to win Audi's first-ever victory in SCCA Trans Am competition.

The 1988 season was not completely dominated by Audi, but it might as well have been. Despite being relatively underpowered and having the largest frontal area of any top contenders, the Audis kept ahead of the pack all year by virtue of traction, preparation and good old racer's luck.

At some events—like Sears Point, where Rohrl gave a thrilling solo performance during a rainy qualifier but had less fun in the actual race—the Audis had to settle for unremarkable finishes. At tracks like Niagara Falls, New York, though, they were untouchable. As the tight, hot street circuit began breaking into marbles, the Audis simply motored on while the competition got beaten up by their own cars. Rohrl took the rest of the field by nearly a lap.

Street courses were particularly kind to the Audis because of the relatively weak surfaces of their pavement. As the tracks deteriorated during the race, other drivers had to struggle to get their 650 bhp down through two overheated tires. The Audi drivers could just stand on the gas and let their "mere" 550 horses take care of themselves. The rain-soaked Mid-Ohio race, too, was a cakewalk for Audi. Not surprisingly, the Quattros could run nearly flat out in the rain while the competition could barely make a circuit without spinning.

But even on long, dry courses like Cleveland, Ohio, and Brainerd, Minnesota, the Audis came out victorious, and that was the point the team had been trying to make. Even on dry tracks where many folks thought all-wheel-drive wouldn't matter, it mattered a great deal.

With the rainy Mid-Ohio win, still three races from the end of the season, Audi had clinched the manufacturer's title in Trans Am. Two races later at Mosport Park, Haywood secured the driver's title as well when both he and his competition failed to

Hans-Joachim Stuck, in his trademark black-and-stars helmet, turned in precise laps in the GTO Quattro. The pipe behind the front wheel is for engine exhaust. *Audi of America*

The IMSA GTO 90 Quattro looks little like the Audi 90 street car, while the production-based Trans Am 200 Quattro was readily identifiable. The four-door body shell remained unique to the series. *Audi of America*

complete the race. His cool, restrained and predictable driving paid off, and the "underpowered and oversized" Audis earned twin titles in their debut year of SCCA's premier series.

Some observers believed that the Trans Am Audis were helped on their way by a good deal of luck. Often, they pointed out, the Audis inherited first place when the previous leaders crashed out of contention. But another element was at work. In the words of Hurley Haywood, one charm of all-wheel-drive is that it "lets you maintain the same characteristics as you started with. If you take cornering speed, we don't necessarily go around a corner any faster, but we can do the same speed lap after lap after lap."

If not out front, then, the Audis were always pushing from behind. A driver could run to the head of the pack, but the red, white and black Quattros would keep coming up in his mirrors like a bad dream. With the other driver forced to push harder and harder, something had to give, and

Located ahead of the front axle and even out beyond the front subframe, the five-cylinder engine placed mass in an inconvenient location. The Trans Am development program and GTO programs centered on moving other masses aft.

Group 44's Winchester, Virginia, shop is scrubbed, spacious and well organized. This is a good place to judge the professionalism of any race effort.

Partially completed GTO race engines arrive at Group 44's headquarters in crates from Germany. Intake tubes and plenum are hand-welded from aluminum stock.

The large reservoir for the dry-sump engine oiling system resides at the far end of GTO chassis, where its mass can counteract some of the engine mass up front.

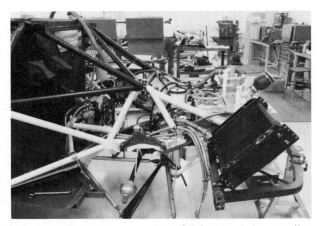

A large cooler at the rear of the GTO chassis keeps oil at respectable temperatures despite track conditions and the tremendous heat put out by the turbocharged engine.

usually it was that driver's concentration or machinery.

Perhaps most telling of all, the SCCA recognized the superiority of the Quattros and levied numerous penalties in an effort to slow them down and tighten up the field. As is their right in Trans Am, the series chiefs demanded additional weight, turbocharger inlet restrictions and even a reduction in tire size. It was all to no avail; Audi and Group 44 simply found more horsepower and stronger brakes to make up for what the SCCA took away.

At the final race in St. Petersburg, Florida—the first in which Audi fielded cars for all three drivers—Rohrl won again despite the host of SCCA-mandated restrictions. He even had time to tangle with another knot of backmarkers in the process. This time, at least, they didn't put him out of the race. There were 155,000 spectators on hand to see

Audi's final burst of intended acceleration in the Trans Am series.

The move to IMSA GTO

In the publicly polite world of racing management, little was said about Audi's move from SCCA Trans Am to the similar but more advanced IMSA GTO series after a single, successful year.

The rumor mill, on the other hand, went wild. Word got around that the SCCA felt Audi had "misled" it about the potential of all-wheel-drive and wanted to get even; some said Bob Tullius wanted to get back to IMSA GTP racing and figured this was a logical first step; others thought Audi was expressing disgust with the SCCA's ever-tightening restrictions.

Perhaps elements of every rumor were true, or perhaps none of them were. But Hurley Haywood, speaking with characteristic candor, saw the situation simply: "SCCA didn't want us there: they

GTO front spindles ride on beautifully formed upper and lower arms, rather than on the MacPherson struts of the Trans Am car. A Heim-jointed steering link is taken up just behind the upper arm.

Burly tubes protect the driver's side door from intrusion in an accident. The same type of driver protection is found on NASCAR racers and other tube-frame cars.

wanted an American series. Our professionalism and our technology was far above the going rate in Trans Am, and they didn't like us.

"Also, the IMSA formula allowed our engineers to be more creative, it gave us more leeway with the cars, and it was a more interesting proposal." In other words, it was a mutually agreeable divorce?

Josef Hoppen—Audi's special vehicles manager and an official company spokesman—issued a less volatile statement: "The rules for Trans Am were such that while we could win, we could not explore the new areas of technology we wanted in order to help develop the engineering for our street cars of the future." New areas like tube frames, carbon fiber brakes and Kevlar body panels? The fans will probably never know just what happened, and the smart racers might never tell.

The fact is that SCCA banned tube-frame cars from having all-wheel-drive for 1989, and that meant Audi wouldn't be able to take the next logical step in the Trans Am Quattro's development. It appears, then, that Audi simply went looking for a home with more liberal rules and more stable restrictions. It found that home in IMSA's GTO series.

There are four levels in the top ranks of IMSA racing. GTU is for the smallest production-based cars, silhouettes built to resemble small machines like the Mazda RX-7 and Chevrolet Beretta. GTO is a similar silhouette series for faster cars like the Nissan 300ZX, Chevrolet Camaro, Mercury Cougar and Audi 90. One step to the side is Camel Light, a prototype series for full-monocoque, mid-engined racers with relatively small engines. Then finally comes Camel GTP, home to no-holds-barred prototypes like Jaguar's XJR-10, Porsche's 962 and the 750 hp Nissan 300ZX GTP.

While IMSA GTO is the series closest to SCCA Trans Am in regulations, GTO cars are allowed many more modifications, and Audi has worked to the full extent of them. Abandoning the Audi 200 platform for the smaller Audi 90, it has built the GTO from a combination of tube-frame, aluminum alloy and carbon fiber components. With the 90's smaller wheelbase, the GTO is 12 centimeters lower than the Trans Am 200 and shorter overall. The wheels are wider, the aerodynamics are better (though the 200's were already excellent), the downforce aids are more advanced and the suspension is completely redone in true race car fashion.

Instead of the MacPherson strut front, trailing link rear suspension setup of the Trans Am and production cars, the GTO-specifications 90 sedan features Formula One style double wishbones all around. The engine went from its single-overhead-cam, two-valve Trans Am layout to a four-valve, double-overhead-cam design, and was bored out an additional 80 cc to 2190 cc. The car was also fit with the same Bosch Motronics engine management system that's seen on the Porsche 962 and championship-winning Porsche-TAG Formula One powerplants. Horsepower was quoted for the start of the year at 620 bhp, and it will no doubt climb throughout the season.

Though the engine is pushed back a bit in relation to the Trans Am car's, it's still hung out ahead of the front wheels. Little was solved in that area, except that by starting from scratch Audi was able to construct the GTO car with this disadvantage in mind, rather than having to compensate for it after the fact. A new gearbox design allows the engine to be relocated slightly rearward, but it still hangs out in the breeze as before.

It didn't have to be that way, but Dieter Basche and his colleagues see the alternative—pushing the engine back behind the front axle and running an external driveshaft forward to a separate front differential—as unnecessary. "We have always said

A relatively empty front engine bay cuts down on forward weight. Sewer-pipe-sized heat shielding keeps intense exhaust heat away from the front differential and driver's compartment.

The compact and tidy rear end of the GTO chassis puts the fuel cell and filler neck at the rear of the car, away from the driver. The oil cooler, fuel cell and oil reservoir all keep their heavy liquids in the tail.

Simple triangulated tube sections make up the bare skeleton of the GTO Quattros. Disassembled to this point, mechanics can change any component in no time flat.

'No, that's too complicated; let's do it like it is on the normal road car.' We have thought about it, sure—it would be an advantage. And maybe one day [if another all-wheel-drive car arrives to compete with the Audi] we may have to go this way. But as long as we can come along with our system which we have in the road cars, we leave it like it is." In other words, Audi sees no need to mess with something that it already knows works.

IMSA GTO offered a stability in regulations that Audi must have found appealing after the frequent penalties levied by the SCCA. The IMSA rulebook merely states that any all-wheel-drive car is subject to a ten percent weight penalty over a two-wheel-drive vehicle of similar displacement. IMSA also reserves the right of a single mid-season adjustment should one car prove too dominant over the rest.

On-track with Audi's GTO racers

By missing the first two events of the year—Daytona and Sebring, two long-distance races that new cars rarely win since they're so hard on machinery—Audi made it difficult to get another set of championships in 1989. That wasn't its inten-

tion (though it hadn't been the idea in Trans Am either and was welcome enough when it came). Rather, the plan for 1989 was merely to get familiar with the more complex and competitive world of IMSA GTO and to continue proving the efficiency of all-wheel-drive in pavement racing. On the pace from their first outing at Miami, by the Watkins Glen race—three from the season's end—the Quattros racked up five wins and four one-two finishes against the V-8 Cougars and GTP-engined Nissan 300ZXs.

Trans Am champ Hurley Haywood describes his experiences with Audi in big-ticket road racing this way: "After [the first test stateside] I knew we had something potentially very good. I was somewhat surprised that we dominated as much as we did, but if you take a look at all the facts, I think you can see why we did.

"It was a very different type of driving style—I was used to ground-effects cars, but once you were able to trust the all-wheel-drive car, you could do things you could never even think of doing with a conventional car. At [Topeka, Kansas, where the GTO Audis finished one-two], I had a tire problem, and I just went sailing off the racetrack at 100 mph in the dirt and gravel. With the all-wheel-drive I just kept my foot in it, didn't let off one single bit, and drove back onto the race course. If I had done that with a conventional car, I would have spun."

Asked about the Quattro's cornering, Haywood adds, "You can go in deeper with AWD because you have more stable braking. Your throttle application is early because you need to get the power down to the ground almost at the middle of the corner. Where in the conventional car you start to apply the power from, say, the middle of the apex, with the [Quattro racers] you start the power almost as you start to turn into the corner.

"With the engine hanging so far forward, you have a car that has a tendency to understeer, but the understeer's now become pretty good. You drive the car with a lot of slippage to it, and once you get used to [that] it is not so bothersome. When I first drove the car I said, 'Holy Christ, I'm going to go off the road!' But when you get used to it, and you understand the characteristics of AWD under

Front differential, transmission and center differential all connect to form a single unit. Oil pickup lines guarantee fresh, cool lubrication at all times.

racing conditions, then you adapt to it and it's no problem. That's why you start so early in the corner with the power."

Where does all-wheel-drive go from here?

As far as Audi was concerned, after the initial and rewarding 1989 GTO season, it seemed likely that the team would return for at least one more

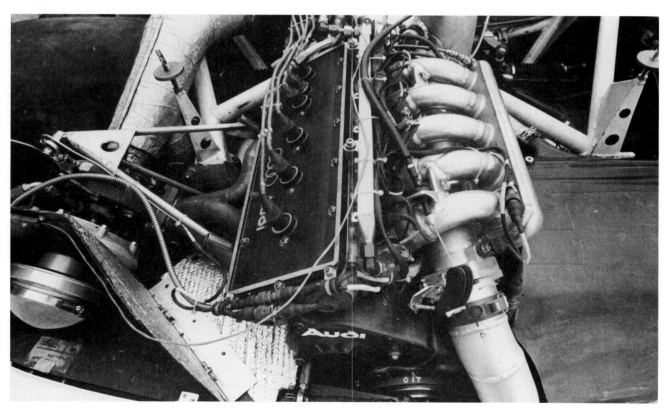

The narrow GTO engine produced 620 hp at the start of the 1989 season, thanks in part to a single KKK turbocharger. A double-overhead-cam, four-valve head didn't hurt either. *Audi of America*

The rear differential is housed inside a husky cast case fit with pickups for oil supplied by its own pump.

The alternator and power steering pump of the GTO Audi are at the back of the car, where they're driven off the driveshaft.

The main driveshaft is made up of superstrong monofilament carbon fiber. Carbon fiber's light weight reduces the rotating mass of the overall driveline package.

The gearbox and limited-slip front and center differentials hide inside the frame of the assembled car; Trans Am rules allowed a six-speed gearbox, but IMSA GTO mandates only five gears.

The dashboard of Audi's GTO racer combines an analog tachometer with multifunction digital gauges. The large knobs to the far right select the function each gauge will monitor.

year in the series. From there, prototype racing would be the next logical step. Such an effort might have appeared in both IMSA GTP and FISA Group C as the series came together in rules and regulations, but an FISA ban on all-wheel-drive has scotched the plan for now.

An Audi Quattro prototype racer would be a fascinating machine. Free to design the layout it wished, it's likely Audi would give us a Quattro that for once didn't win *despite* its reliance on production technology. An all-wheel-drive prototype ought to be fast, and Audi must realize that if it doesn't do it, someone else will. Still, would the excellent traction and cornering of a ground-effects prototype chassis negate some of the traction advantage of all-wheel-drive? We'll just have to wait and see.

Tommy and Bobby Archer showed some interest in pushing their own all-wheel-drive involve-

Shorn of Kevlar body panels, the tube-frame GTO car looks downright ungainly. (The mechanics have stepped away for the photo; they didn't just abandon their tools on the floor.)

Adjustable gas pressure coil-over Koni spring and shock units top off the GTO's independent rear suspension. Unlike the Trans Am car, the GTO uses no major production parts.

ment further until the SCCA came down on all-wheel-drive in Trans Am. Their idea for an all-wheel-drive Talon silhouette Trans Am car was cut off in the planning stages. "The SCCA...banned the car," Bobby says, "before they ever saw it."

As the nineties began, it was generally agreed that while all-wheel-drive had finally been proven in road racing, a hurried rush to embrace the technology was still some time off. No doubt more and more racers are going to use it, but few foresee an overnight revolution. According to Brian Berthold, "The fact of the matter is that if everybody else ran four-wheel-drive, because of their lack of development and experience with it, they would not be as competitive as they would without it. It simply took too much work [for Audi to] get it where it is." Berthold explains further by pointing out that just adding all-wheel-drive hardware isn't enough. The hardware has to be tested and developed extensively before it will perform the way a racing team wants it to. "Development is the key," says Berthold.

There's no doubt that almost all the recent development in serious all-wheel-drive racing applications has come from Audi. The people from

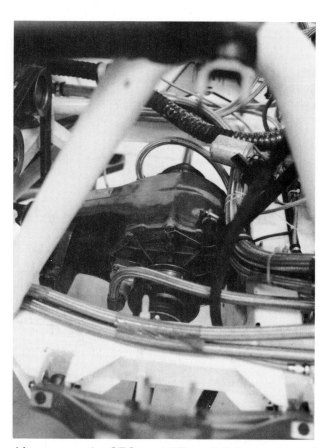

After a race, the GTO rear differential shows signs of weeping oil around seals and the vent tube. Safety wire on case bolts and zip ties on oil lines keep everything in place.

101

Passenger's side of the GTO tube-frame lacks heavy anti-intrusion bars. KKK turbocharger sits at far front right of vehicle and uses a GTP-style air restrictor while racing.

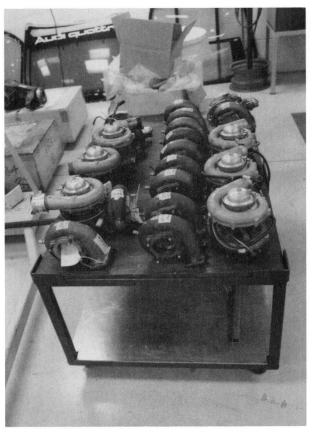

KKK turbos await the call of duty at Group 44 headquarters. Audi has been a leader in turbocharging since the 5000 Turbo street sedan. Turbos also helped Audi's Quattro rally cars make their considerable horsepower.

Ingolstadt, West Germany, definitely have a jump on the competition in terms of experience, but they know they're not the only team with excellent engineers.

When a technology becomes necessary for winning races, other teams have a way of learning it very quickly indeed. And if winning means intensive research and development, that's simply what the good teams will have to do.

For a while yet, people will keep pointing to Audi and the Archers as the best-organized, best-financed and best-engineered outfits in their respective circuits, and using those qualities as an excuse to demean the role of all-wheel-drive in their success. But for both teams to have won major titles with notably less power than their rivals—in cars that, at least on paper, should not have been winning—then something more than just superior organization is at work.

The formula for choosing a winning car on paper may soon need a new factor added to it. That factor is all-wheel-drive. Sooner or later, it's coming to a racetrack near you.

Chapter 6

Taking it to the Street

If you're interested in all-wheel-drive vehicles, you're obviously something of a car enthusiast and probably already a pretty decent driver. So this chapter isn't going to start from scratch and give you the lessons of a Bob Bondurant class. Instead, it will address some special points of buying and living with an all-wheel-drive car.

The most important points covered here are safety-related. Why stress things like car control, poor-surface driving and skid recovery when all-wheel-drive automobiles are more stable, more predictable and just plain safer than their two-wheel-drive counterparts?

For just those reasons. Because of the superior traction and roadholding of all-wheel-drive, chances are you'll wind up going faster in the dry, going faster in the wet and just plain going out in more kinds of weather than you would in a conventional automobile. And while all-wheel-drive performs much better under poor driving conditions—and yes, in good conditions, too—no automobile is completely stable.

You should take advantage of all-wheel-drive's superior roadholding ability, but at the same time realize that all-wheel-drive doesn't make you invincible. The faster you go in snow, rain and so on, the

A natural combination: Subaru all-wheel-drive and the United States Ski Team. Snow is by no means the only condition in which all-wheel-drive is an advantage, but it's the one many people think of first. *Subaru of America*

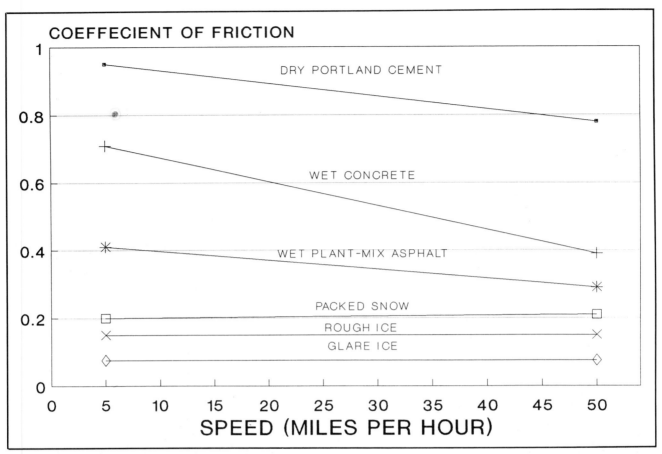

The coefficient of friction of most road surfaces drops off with increased speed, but with ice and snow it starts off low and stays that way. *Tom Smitham Graphics*

more alert and aware you need to be. And if you are going to use your car harder and more often than a two-wheel-drive owner, you must be better at avoiding accidents, as well.

Purchasing an all-wheel-drive

Any big purchase like an all-wheel-drive vehicle is going to be a decision between what you need, want and can afford. If you have money sitting around in sacks, you can just trot out and buy the ultimate all-wheel-drive machine. If you're like the rest of us, though, you have to consider the options and what each one is worth to you.

The value of any all-wheel-drive system depends on the car's intended use. If you're buying all-wheel-drive for its added high-speed stability and cornering, you'd better set yourself up in a car with a full-time drive system. The only way to be guaranteed all-wheel-drive performance at all times is to get a system that's never disconnected.

If you're after safer driving in the long winter months only, an on-demand system might work fine for you. Automatic on-demand systems are great in that they don't require any added driver input, but real hot-shoes might find their on-again, off-again behavior troubling.

And if you're after all-wheel-drive only for those rare vacations off the road, you might be perfectly happy with a part-time system you can throw in just when the going gets really rough. If you're in this group, though, remember that for the money you spend on all-wheel-drive, you could pay to have chains put on your two-wheel-drive car a whole lot of times. You could also spend a lot fixing dented body panels, broken bones and so on. If you don't get around to chaining up at the right time, so this might not be such a bargain in the long run. Also consider that all-wheel-drive may let you take a trip you'd otherwise cancel owing to lousy weather.

If performance driving is your goal, you also have to choose between a front and a rear primary axle. People used to front-drive cars will be more comfortable in a car with a primary front axle (fifty-fifty torque split), while those used to power sliding strong rear-drive cars will enjoy a rear torque bias.

Variable-ratio cars like the Porsche Carrera 4 will make the choice for you, and they generally get it right. The Carrera 4 uses an interactive system in which an onboard computer monitors wheel speeds from the ABS sensors, correlates the information with throttle position and other input, and adjusts the multiplate center and rear differentials accordingly.

As mentioned elsewhere, you handle an all-wheel-drive vehicle very much the same as you do a front- or rear-drive car, depending on its torque split. All-wheel-drive will allow you to get away with many brazen maneuvers you normally couldn't do with two-wheel-drive, but the mechanics of driving both types are the same.

Differentials are another option that you have to take into consideration. Will it bother you to have to manually lock your differentials in rough going, and unlock them again once you're through the slop? Bear in mind that manually locked differ-

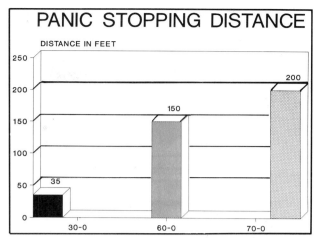

Stopping distances increase exponentially with speed. It can take just 30 feet to stop from 30 to 0 mph, but more than 50 feet to slow down from 70 to 60 mph. *Tom Smitham Graphics*

The Quattro driveline and fast German roads make a rewarding combination in the rain, but no car is invincible. Whatever speed this driver is traveling, he could go faster in the dry. *Audi of America*

A clear and present danger of hydroplaning. Deep water and the light front end of the mid-engined RS200 don't help matters, but rally drivers always treat standing water with the greatest respect. *Ford of Europe*

entials mean deactivated ABS, too. If either of these is a problem, you'll have to look for a vehicle with some sort of multi-clutch, viscous or Torsen differentials, and you can expect to pay a little extra for them.

Traction of various road surfaces

The Society of Automotive Engineers (SAE) has spent loads of time testing and quantifying the coefficients of friction for various road surfaces under different weather conditions. Not surpris-ingly, it has found that the traction offered by any type of road varies greatly, not only by its composition but by the speed the vehicle is traveling and whether or not the surface it is traveling over is dry or wet.

The coefficient of friction of fresh, dry Portland cement is about 0.95 k. But that's only true at 5 mph—it falls off to about 0.78 k at 50 mph. On wet Portland cement, the coefficient begins at about 0.82 k and falls to about 0.75 k at 50 mph.

Portland cement in good condition offers excellent traction, but most surfaces do not. Plain wet concrete's figures are 0.71 k at 5 mph but just 0.39 k at 40 mph. Wet plant mix asphalt starts with 0.41 k at 5 mph and falls to less than 0.30 k at 50 mph. Further down the scale, packed snow offers a coefficient of friction lower than 0.20 k at 10 mph, which *increases* slightly to about 0.21 k at 50 mph. Rough ice, depending on its surface condition, stays between 0.10 k and 0.20 k without changing noticeably at legal speeds, and glare ice stays constant at around 0.05 k to 0.1 k.

Before you go speeding up on packed snow or ice, however, remember that while the force of gravity—hence the coefficient of friction—remains constant with speed, inertia *increases exponentially* with speed. In other words, while snow and ice will accept a given amount of force regardless of how fast the car is traveling, the force needed to

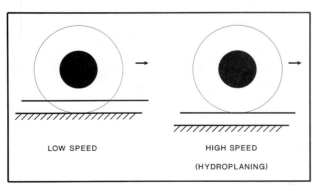

LOW SPEED

HIGH SPEED

(HYDROPLANING)

At low speed, the tire is pressed through an interfering layer of water and makes contact with road. When the tire is traveling faster, water can't escape and forms a virtually frictionless layer. *Tom Smitham Graphics*

As the tire rolls over the road, thin snow melts and the car behaves as if it were on a wet road. Earlier cars have already melted tracks in this section of English road. *Ford of Europe*

Light snow over packed snow creates a treacherously slick roadway, with predictable results. All-wheel-drive's grip can't overcome the excessive speed of Kalle Grundel's Ford. *Ford of Europe*

Pneumatic suspension gives the Subaru sedan high ground clearance, but a tread clogged with snow might stop this car in its tracks. A little more snow and chains would be in order. *Subaru of America*

change the car's direction increases dramatically as speed rises. Driving more slowly is the only way to remain in control.

The increase of inertia with speed also affects stopping distances, as you may remember from driver education class. While it might take just thirty or forty feet for the alert driver to stop from 30 mph on good pavement, the distance increases to more than 200 feet at 70 mph. These figures go up dramatically as the quality of the road surface goes down, and they don't include driver reaction time.

The implications are clear. As speed increases, the ability to stop in time to avoid an obstacle in front of you decreases not linearly but exponentially. Any condition that decreases visibility, such as fog, rain or darkness, warrants a decrease in speed. There are some things you can do to better your chances in low-visibility driving, however, and maybe even get home a little faster.

Driving in fog

In fog, at least, the road surface is generally clean and offers decent grip. Visibility, not surface condition, is usually the main concern. If you can follow the lights of a vehicle in front of you while still retaining enough room to brake should that vehicle stop, by all means do it. If not, it's better to drop back and go it alone. It's not unusual for a whole slew of cars to pile into each other because

each driver was following the car in front and all were going too quickly. (More than 200 vehicles once piled up this way on U.S. Interstate 5 one foggy California morning.) Fog lights are a good idea, because their beams are cut off below the level that would bounce back into the driver's eyes. Low beams are preferable to high beams in fog for the same reason.

Many all-wheel-drive automobiles come with fog lights standard, but Cibie, Hella and other aftermarket manufacturers make add-on lights that can work wonders. Since your all-wheel-drive hardware won't help you too much in fog, they're worth checking out.

Driving in rain

Rain combines both visibility and surface problems, and here all-wheel-drive *can* help you out. As in any low-grip situation, you'll be able to stop and go more quickly with all-wheel-drive than without it, but you still can't go as fast as you could on dry pavement.

The coefficient of friction of a paved road decreases dramatically in the first fifteen minutes of a rainstorm as grease, rubber and general crud are washed from the pavement and float to the surface of the water. Anyone who's had the pleasure of driving on a Los Angeles freeway during the first rain of autumn knows how treacherous this can be; the highway becomes as slick as ice. After the rain washes the foreign matter away—it often takes no more than fifteen minutes of hard rain—grip increases again to a steady level below that of dry pavement.

The real danger of driving in rain is hydroplaning. Over a wet road, the tread grooves of a tire are supposed to carry water away from the space between the tire and the roadway. If no provisions were made to move the water out, a standing layer would form between the tire and the ground, effectively creating a hydraulic bearing. This is hydroplaning, and anyone who's experienced it knows it's no fun at all.

Hydroplaning is a function of speed, water depth, road surface, tread design and tread depth. All but the water depth and road surface are controllable by the driver. The faster a vehicle is traveling, the less time there is for water to be moved from beneath the tread, so hydroplaning occurs sooner at higher speeds.

The design of the tread is critical to how quickly water is moved from the contact patch, but tires that move lots of water tend to be noisy and don't necessarily handle well in the dry. A choice between wet- and dry-weather performance often has to be made when buying tires. Similarly, wide tires that provide lots of grip under good conditions can lead to hydroplaning in the rain. They spread out the weight of the car across a larger surface and force the water to travel a longer distance before it is channeled out from under the tires.

Finally, the deeper the tread, the more resistant a tire is to hydroplaning. Deep tread also means tire squirm and imprecise handling in the dry, though, so there's a decision to be made here, too.

Michelin and Pirelli offer tires specifically designed for all-wheel-drive performance applications. The fairly open and aggressive tread pattern of the Michelin MXT4 resists clogging with snow and mud. *Michelin Tire Company*

Ford's only all-wheel-drive automobile in 1989, the Tempo/Topaz, demonstrates all-wheel-drive's effectiveness on ice at the Ford/Michelin Ice Driving School in Steamboat Springs, Colorado.

109

While drivers can't control the road surface or puddles, they can at least recognize which types of both are most likely to cause problems. Deep puddles are naturally the ones to watch out for, but even standing water a millimeter deep—the amount on a well-drained highway during a heavy rain—will cause hydroplaning under some circumstances. Roads with crowns, rough surfaces or grooves are much more likely to be safe at higher speeds than those without them.

When hydroplaning occurs on a front-wheel-drive car, all control is lost. The front tires leave the road surface and the driver can't brake, steer or even accelerate. With a rear-wheel-drive, the driver has the minimal consolation that at least one aspect of the car, acceleration, is still under control. Because the front tires generally clear a path through the water, it's very rare for both front and rear tires to hydroplane.

All-wheel-drive, front-wheel-drive or rear-wheel-drive, there's not a lot you can do when your steering goes limp owing to hydroplaning. What you *shouldn't* do is get on the brakes. Gently close the throttle off, and with an all-wheel-drive vehicle it's also a good idea to declutch, letting the front wheels rotate more readily. After that, hold the wheel straight ahead and ride out the longest few seconds of your life.

Driving in snow

Snow offers a whole range of driving conditions. Light snow can be treated just like rain, as the tire rolling across the road immediately melts the covering of snow and turns it to water. Light snow over packed snow merely puts the lubrication of water over the already slick surface of packed snow and reduces traction to almost zero. Packed snow reacts almost like ice, but the ruts left from cars that have gone before will often pull a vehicle along like railroad tracks. If the ruts go where you want them to, fine; if not, you have to plan far ahead on where and how you intend to break free and go your own way.

Starting off in any sort of snow is usually easier in second or even third gear, where using less torque is helpful in avoiding wheel spin. In fact, any maneuver that results in less force being transmitted to the road is desirable. Smoothness is the key to driving in any situation, but on snow it's absolutely essential. Gentle, slow inputs and a conservation of speed can get you through many situations that will stop most two-wheel-drive cars cold.

The best way to deal with snow is to make sure you start out with the right tires. Narrow tires that concentrate lots of weight on a relatively small contact patch go a long way toward making winter driving livable, and their tread pattern should be relatively open to avoid getting clogged with ice. Four snow tires for your all-wheel-drive may seem expensive at first, but they can make driving a whole lot easier and might pay for themselves by helping avoid a fender bender.

Studded tires are great if you're fortunate enough to live where they're legal, but most of us have to resort to chains when the going gets *really* bad. Chains are often a necessary evil: Putting them

The added downforce of spoilers aids high-speed traction dramatically. The Audi Quattro Coupe's lightly loaded rear end gets planted by a large fiberglass duck tail. *Audi of America*

A few vehicles allow drivers to shut off their antilock braking systems. Smart drivers leave ABS on except in rare situations. *Audi of America*

Mud can often be treated like snow, and rear-wheel-drive cars have serious traction problems in both. This Cortina driver has found an entertaining, if dangerous, way of adding traction. *Ford of Europe*

Malcolm Wilson may be able to recover from a slide this serious, but most drivers can't. The real danger is that the rear wheel will dig in and lever the MG 6R4 onto its side. *Kermish-Geylin Public Relations*

Most drivers don't take a light layer of water like this seriously, but if it has floated oil and rubber to the surface of the road it can be treacherous. *Audi of America*

In extremely slick conditions, the driver of the Audi Quattro Coupe can lock the rear differential for added traction. The center differential is a self-regulating Gleason Torsen. *Audi of America*

on is a great way to get frostbite and skinned knuckles, and one of the few ways yet invented to run yourself over without someone else's help.

You can put off resorting to chains much longer with all-wheel-drive than with two-wheel-drive, and you might well make it through the winter without ever bringing the infernal devices out of the trunk. Most highway departments allow all-wheel-drive vehicles to pass through chain control stations for two-wheel-drive vehicles until the weather becomes very bad indeed. In deep snow, however, you're not going to have much alternative. The ultimate responsibility rests with you—not some person in an orange vest—so you have to make the call.

The main benefit of chains is simple: They give the car's tires something that can push against deep snow, which slick rubber can't do. An additional benefit is that they also limit speed in conditions where speed should indeed be kept to a minimum. In an all-wheel-drive car, it's easy to become lulled by the sure-footedness of your automobile in snow and to wind up traveling too quickly. Driving more than about 25 or 30 mph with chains not only destroys the chains themselves, it sets up vibrations so severe that most drivers acquiesce and slow down before their fillings fall in their laps.

If you do have to resort to chains in your all-wheel-drive, the natural question is Where do I put them? On all four wheels, of course. Won't *that* be fun?

In snow, your all-wheel-drive car's acceleration and response will be significantly higher than a

Sand presents the danger of getting stuck but less danger of a serious accident. The top few inches of sand are the hardest, so going fast enough not to break through while still staying in control is the key. *Audi of America*

two-wheel-drive car's, but your braking will be marginally better at best. Without chains your braking will in fact be much worse than that of a chained two-wheel-drive. So even though you can leave those chained-up Oldsmobiles eating salt spray, it's a bad idea.

Remember, too, that ABS gives you *longer* stopping distances over light snow and gravel than you would have in a full-lock slide, because snow tends to ramp up ahead of the tires without ABS. For mechanical reasons, ABS is often disengaged when the center differential is locked, and you can take advantage of this in some cars. But remember that intentionally locking your brakes is never to be taken lightly. You might stop a little sooner, but you're giving up all steering ability to do it.

For the truly advanced driver, rally champ John Buffum mentions that the handbrake can be used to intentionally start a slide if necessary. Since most of us will wind up in more trouble than we started with this way, this is a trick for seasoned pros only.

This Audi driver has managed to get his 4000CS Quattro to oversteer on a gravel road. Gravel can be as treacherous as snow at high speed. *Audi of America*

Driving on ice

Ice is probably the worst driving surface in the world. Ice simply exaggerates the handling characteristics of an automobile, bringing oversteer and understeer to levels that would be downright funny if you weren't sliding toward the boss' BMW when you discovered them.

If not in front of the boss' BMW, you're most likely to encounter ice in shaded parts of the road and on bridges. (Bridges freeze soonest in cold weather, because air is circulating all around them instead of just across the top of the road surface.) Since corners often get the most run-off water from a snow bank, as soon as the temperature starts to drop, all corners should be suspect even if the rest of the road is clear. Nighttime is the worst time for ice, but even a seemingly sunny winter day can lay lots of slick stuff across the road.

Another place to be careful with ice is *underneath* a seemingly innocent layer of snow. Melting snow provides a constant flow of water that can freeze long before the weather makes ice a problem on clear roads.

With the minimal grip offered by ice, the conservation of traction that all performance driving incorporates becomes a critical exercise. The slightest weight shift due to acceleration or deceleration can unload one end of even an all-wheel-drive automobile and result in a slide. Regaining control on ice is perhaps the trickiest maneuver a driver will ever have to master, so a slide should be avoided in the first place whenever possible.

Because ice is such treacherous stuff, a number of schools have opened up in Europe to teach drivers how to handle themselves on frozen roads. So far there's just one such school in America, the Ford/Michelin Ice Driving School in Steamboat Springs, Colorado. French rally ace Jean-Paul Luc comes to Steamboat Springs every winter and teaches state troopers, emergency drivers and even average Joes like you and me how to increase their chances of having a dent-free ski season.

For a reasonable sum and one afternoon of your life, Luc will teach you how to deal with the frozen roads of winter. The school uses all Fords, of course, and makes sure that there are representatives of front-drive, rear-drive and all-wheel-drive vehicles at each session.

When asked how to survive on ice, Luc says, "The first step is to reduce your speed to the driving conditions. If you drive too fast you will not be able to avoid a spin. That's the main rule. I remember I was in Cleveland and it was snowing, and everybody was driving the same speed as [they would] on a dry road. Because of that, a car began to spin in front of me—in such a case you have no chance

Locked wheels on gravel build up a wedge of stones in front of tires, decreasing stopping distance. This driver would lose steering control, though, so would it be worth it? *Audi of America*

to *avoid* a skid." Reducing your speed to driving conditions on ice can sometimes slow you to a crawl, but crawling is still faster than waiting for a tow truck to pull you out of a snow bank.

But the only way most of us learn what a safe speed on ice might be is to exceed that speed a few times and deal with the consequences. Sooner or later, even the best driver is going to get into a situation on ice where the car enters a skid. Once you're in the soup, the trick is how quickly you can get back out.

Jean-Paul Luc's advice once you're in trouble is this: "The first thing to do is react very quickly. And to react you have first to turn into the skid." In other words, steer toward where you *want* the car to point, not where the car *is* pointing. "Then you need to avoid panicking and doing the wrong thing, which will put you in a critical situation. Many times, the average driver goes on the brake as soon as the spin starts—that's the wrong thing. Avoid any weight transfer that can imbalance the car. In many cases, if you also take your foot immediately

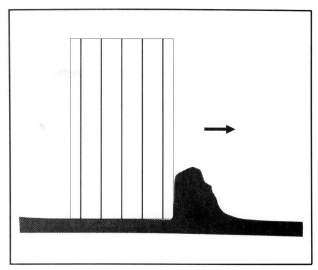

A wheel sliding perpendicular to the vehicle's direction of travel can build up a wedge of soft dirt and lever the car onto its side. Steering into the skid prevents this at least on the front wheels. *Tom Smitham Graphics*

Corners, shaded spots and roadside parking areas freeze first when temperature drops. The Toyota Corolla

All-Trac wagon shouldn't have any trouble pulling out, but a rear-wheel-drive car might. *Toyota*

The Subaru Sedan gives buyers a choice of on-demand or full-time all-wheel-drive. Extras like turbocharging and adjustable pneumatic suspension must be considered individually. *Subaru of America*

Five-speed manual or four-speed automatic both come with full-time all-wheel-drive and ABS on the BMW 325iX. Two- and four-door body shells add another variable. *BMW of North America*

The sweet first-generation Celica All-Trac Turbo throws another alternative into the all-wheel-drive buying mix: used cars. *Toyota*

The relatively inexpensive Toyota Corolla All-Trac sedan offers low-cost full-time all-wheel-drive, room for five and modest but sufficient acceleration. *Toyota*

Honda calls it Real Time 4WD, a euphemism for automatic on-demand all-wheel-drive. Viscous coupling automatically engages the rear axle when added traction is needed. *Honda North America*

off the gas, you are going to provoke a weight transfer which is going to imbalance the car. You will unload the rear wheels, which is the worst thing to do when a car starts to oversteer.

"Turn into the skid, do not go onto the brake, do not lift off the accelerator abruptly, just *try to keep the same speed*. If you do so in the right way, you'll avoid a counterskid. A counterskid is the rebound motion of the suspension, which sends you to one side and then the other. That is very, very difficult to control—often it's impossible."

A counterskid—Americans usually call it overcorrection—can occur in any skid-correction maneuver. It's dangerous on dry pavement but even more treacherous on ice, where every skid is grossly amplified. An overcorrected skid throws the car's own springs into the equation against you, so it's often much more serious than the first skid.

But Luc feels that all-wheel-drive goes a long way toward taming ice's nastier side. "In all-wheel-drive you have twenty-five percent of the power on each wheel instead of fifty percent on two, and because of that you have less chance to spin the wheels. By having less chance to spin the wheels, you have more chance to keep control of the car.

"Really, I'm very positive on all-wheel-drive. We teach some advanced driving techniques at the school in which we demonstrate that with all-wheel-drive, using some of the right techniques, you can really keep control of the vehicle. It's

incredible how you can keep control of the vehicle—I love all-wheel-drive."

But do you have to adapt your ice-driving techniques to having power at both ends of the car? Luc answers: "Not really. The main difference is that everything is easier to do. On a rear-wheel-drive vehicle you can oversteer very easily, and because of that you can lose control very easily. On the front-wheel-drive car, when the car understeers, if you stay a little bit too much on the accelerator, you have some chance of spinning the front tires. All-wheel-drive is more neutral, and because it's more neutral it's easier to control."

If there is a disadvantage to all-wheel-drive on ice, it's the old devil of overconfidence. Luc says, "You drive all-wheel-drive and you see how easily you can move your car. You really feel more confident with this kind of car. You can move it more easily, and also you can slow down more easily. When you drive a rear-wheel-drive vehicle, and you have a lot of problems just moving your car from the parking lot, you're not very confident with your vehicle." Luc goes on to say that higher confidence can lead to higher speeds and therefore an increased likelihood of getting into trouble. "Because of that you have to think of the next corner; you have to think how you have to slow down."

If you're lucky enough to have ABS on your all-wheel-drive vehicle, you'll probably have much less trouble on ice than will drivers who are doing without. Luc points out that locking a wheel on ice is a sure-fire way to start a skid, and ABS can haul a car down on ice without any locking whatsoever.

If you're stuck without ABS, ice is a good place to start using cadence braking, the technique drivers always talk about but rarely bother to use. Cadence braking is essentially the human imitation of ABS. By pumping the brakes rapidly, you bring the car to the point of, but not into, imminent lockup over and over again. Since normal hard braking on ice is virtually impossible—and a great way to get into a nasty, uncontrollable slide—you've got nothing to lose with cadence braking, so you might as well give it a try.

More on bad-weather driving

If you manage to get your all-wheel-drive honestly and truly stuck, after a suitable period of feeling embarrassed you'll have to figure out a way to extricate the vehicle. The general rule is that what works for two-wheel-drive cars will work even better for all-wheel-drives. Whatever it takes to get out of snow, incidentally, will get you out of mud as well.

Once the differentials are locked and all-wheel-drive is engaged (in part-time or on-demand systems), if you're still stuck it's time to get some friction between your car and the ground. Got any rubber floor mats? Sacrifice them between the tire and the snow. If not, try evergreen branches, blankets, the famous bag of kitty litter or whatever happens to fall to hand.

Often, rocking the vehicle back and forth by alternating between second and reverse gears will build up enough momentum to get you unstuck. Willing helpers to push and rock the car are handy, too, if you're lucky enough to have some around.

A tune-up in the fall is a good idea whether your car is due for one or not. Cold mornings and thin mountain air play havoc with many cars' intake and ignition systems, so having them in top condition beforehand is always advisable. The same goes for the battery, as it will only offer about sixty-five percent of its warm-weather cranking power at 32 degrees Fahrenheit.

Windshield wipers and washers are a convenience in the summer, a necessity in the winter. Make sure you've got some washer-fluid antifreeze in the reservoir (engine antifreeze will ruin your paint job) and keep the wiper blades in good shape. In addition, make sure you've got a flashlight *that works*, some jumper cables, a complete set of tire chains and a first aid kit in the trunk. If you're going to be spending a considerable amount of time in poor conditions, you might want to add a citizens

Soft dirt usually has enough support underneath to allow reasonably spirited driving, but hardpan or rocks below the surface can change traction conditions immediately. *Subaru of America*

119

This driver kicks out the rear end of a rear-wheel-drive car with throttle, saturating the rear tires and causing a rear-end slide. All-wheel-drive with rear torque bias can be driven the same way. *Ford of Europe*

Throttle application in front drive or front-biased all-wheel-drive causes the front tires to saturate and slide. *Subaru of America*

band radio or cellular telephone to your all-wheel-drive as well. If you get stuck, the hike back to the nearest telephone to call for help can be the most dangerous part of your journey.

As for the surface of the car, winter is the toughest season. One school of thought says the best way to protect your car from the ravages of salt and water is to coat it in a thick layer of mud in the fall and then let it accumulate grime through the rest of the year. The only good thing about this idea is that it saves you the trouble of washing your car. Other than that, it's complete hokum.

The only ways to protect your vehicle in the winter are storing it in a garage, having it rust-proofed by the dealer or factory, giving it a good coat of wax and *keeping it clean*. Keeping it clean means regularly washing the top, bottom and particularly the inside of the wheelwells, where salt and crud accumulate fast. Self-service car washes work well since they can give you more water pressure than can most garden hoses, and the washing wands are angled to rinse out hard-to-get spots.

Avoiding and minimizing accidents

If you're always driving within your vision range and within your stopping range, you should never have a collision, right? Wrong.

Situations simply occur that even the best driver can't foresee, and at one time or another it will probably happen to you. Within a split second, you'll have to evaluate the situation and take the proper steps to avoid it. In most cases, good, correct reactions will get you out of nearly the deepest of troubles. Sometimes, however, impact is unavoidable.

If possible, the idea is always to avoid contact. The key here is being alert. Make a conscious effort to be aware of traffic all around if you hope to avoid a distinctly checkered driving record. Just remember that if you're not paying attention, you're going to hit stuff. Guaranteed.

It's important to be able to recognize safe options in an emergency and to have the nerve to take them immediately. A safe option is one that won't just buy you time but will minimize or avoid

In this severe example of weight shift, cornering hard and accelerating shift enough weight to the right and rear that the front left tire leaves the ground. The late Jim Clark is at the wheel. *Ford of Europe*

The Volkswagen Vanagon Syncro's viscous transmission relegates the front axle to on-demand status. When the rear axle slips, the difference in over-the-road wheel speeds hooks up the viscous transmission. *Volkswagen of America*

contact completely. Many people avoid hitting a parked vehicle in front of them by swerving into the path of an oncoming truck farther down the road. Not smart. Better to swerve onto the highway shoulder on the right.

Only experience can tell you how hard you can wrench on the wheel or the brakes and remain in control of your vehicle, and remaining in control is what you'll need to do. Many driving schools now feature accident simulators, usually in the form of one lane that branches into three controlled by traffic lights. The drill is simple: You speed down the single lane, and at the last moment one light at the branch goes green and the other two go red. You have to get into the green lane quickly and not wander into either red one. It's hard to do, but because you're expecting it it's still easier than reacting to a real-life emergency.

Most people's tendency is to feed in too much input too quickly in an emergency, but better that than doing too little too late. It's possible to regain control after an overdone avoidance maneuver, but getting in an accident can't be undone. Rod Hall, America's premier off-road racer, has a simple secret to avoiding accidents: "Never give up," he says. "Keep steering until you stop—or hit something."

If contact is unavoidable, it's in your best interest to hit something soft and forgiving. Many drivers will merely get on the brakes and hope for the best if contact is unavoidable, but if you can go through a fence, a hedge, a thin tree or a breakaway light pole instead of into a stopped vehicle, do it—but watch out for pedestrians; hedges, trees and fences are usually on their part of the road.

The good news

Now that all the doom and gloom of our safety discussion is out of the way, take heart: Auto experts agree that an all-wheel-drive vehicle can save you in maneuvers that would leave two-wheel-drive cars sailing for the wrong side of the road.

Road-racing champ Hurley Haywood says: "I run a driving school for Porsche customers in Savannah, and have done some back-to-back tests with Porsches and Audis. Most people are good drivers when everything is going just according to plan, but when an emergency happens, they forget everything, and they get themselves into bad situations. They just don't know how to control their car under extreme driving conditions. With AWD you can avoid accidents much easier than with RWD.

"It happens many times under normal street driving conditions where you're going from dry to wet and back to dry again. If you're hustling through that corner and suddenly you get into a wet situation—or ice or sand—the car wants to understeer. Then you suddenly get off the gas, and the car wants to spin. You could easily spin into oncoming traffic, a guardrail, any number of ways.

"In an all-wheel-drive car in that same condition—even if you suddenly let off the gas or hit the brakes—maintains its integrity and keeps going straight. Just because it has all-wheel-drive. It simply gives you better stability under any kind of driving condition."

Or take the opinion of Group 44 racing engineer Brian Berthold: "If you drive at thirty percent of the car's capabilities you could have one-wheel-drive. You'd never notice the difference. But the closer you get to the limit, the all-wheel-drive gives you that much more margin of control. Whatever those limits are or however they're created, whether you drive hard or in rain or snow or ice, the AWD simply leaves you that much more margin—whether it be for error, unexpected road conditions, or because you're driving at nine-tenths and a cat runs out in the road."

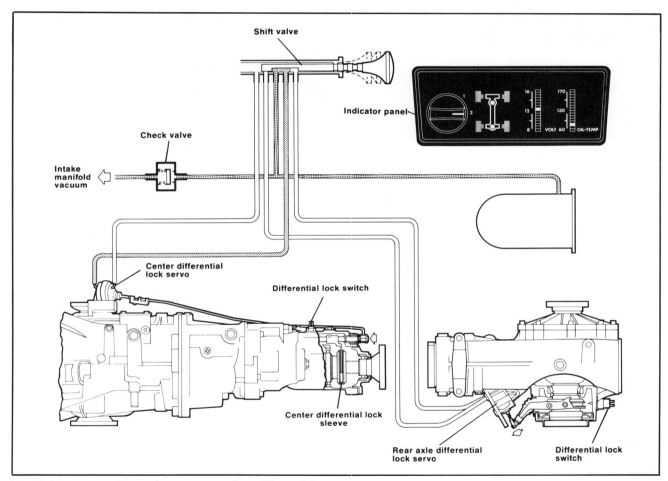

Manifold vacuum is used to engage the differential locks in one Audi system. The driver must select the proper setting for conditions. *Audi of America*

Chapter 7

Futurelook

Automobile manufacturers have used show cars to get the public primed and ready for coming ideas almost since the beginning of the industry. On occasion they'll even send a show car straight into production, as they did with the Chevrolet Corvette, Porsche 959 and Pontiac Trans Sport.

More often, however, show cars merely provide glimpses of some of the things buyers can expect in years to come. Some of the design or engineering of almost every dream car will appear on the showroom floor eventually. Though most show vehicles provide just an idealized version of future production cars, the technologies used in building them should be taken seriously. If a manufacturer is looking toward a new idea, you can almost bet it will appear at an auto show long before it comes to the world's garages.

Through the eighties, an almost de rigueur feature of the world's most exciting show vehicles was all-wheel-drive. In addition to all-wheel-drive's simple exotic appeal, the high performance—or at least high theoretical performance—of these future-minded specials made all-wheel-drive nearly mandatory. The show cars of recent years demonstrate the position all-wheel-drive will hold tomorrow in the real world of production cars.

Porsche 959

In 1981, Porsche showed a smart little convertible based on the 911 platform and powered by a 3.3 liter turbocharged engine. All well, good and relatively tame stuff.

But seemingly at the last minute it threw all-wheel-drive into the package. The conversion basi-

The Gruppe B show car was virtually identical in body design to later 959 street cars. Still, it took three years of research and development to get the car into production. *Road & Track*

The all-wheel-drive 911 for the 1984 Paris-Dakar stressed simplicity and rugged construction over high technology. A normally aspirated, air-cooled Carrera engine supplied the power. *Road & Track*

High-tech 959 rally cars made short work of the 1986 Paris-Dakar, handily coming in first and second. By this time, excitement over the impending street car reached a fever pitch. *Road & Track*

The Porsche 959 racing model, dubbed the 961, appeared at Daytona's IMSA finale in 1986. It was one of the 959's rare appearances in any form on America's side of the Atlantic. *Road & Track*

cally added a 924 transaxle, turned around 180 degrees, to the front of the car. A simple transfer box and driveshaft meant the all-wheel-drive wasn't very sophisticated. Porsche's executives just stood by and smiled when they were quizzed on the meaning of this all-wheel-drive 911, and people soon forgot about it.

At the same Frankfurt, Germany, auto show two years later the concept was back—this time in force. The show car was called simply Gruppe B, and in production it promised to be the most exciting Porsche road car ever. If, indeed, production was really planned. In 1983, no one knew. Today we do.

The Gruppe B was based loosely on the 911, but almost none of its parts interchanged with those of its less exotic little sibling. Body panels that harked back to the elderly 911 shape were in fact made of Kevlar and formed so organically that despite its aerodynamic aids the coefficient of drag was just .32.

The familiar two-plus-two 911 seating also remained, as did the flat six, rear-mounted engine, but there the similarities pretty much ended. The show car boasted a six-speed gearbox ahead of the rear axle (the same gearbox eventually used in the Audi Sport Quattro S1), electronically controlled multi-plate clutches for the center and rear differentials, torque tube drive to the front wheels and an Indy-bred, quad-cam, turbocharged engine with about 400 hp on tap.

Torque distribution was variable from 100 percent rear drive to fifty percent front, fifty percent rear, controlled both automatically and by the driver through a stalk on the steering column. It was an interactive system: The driver selected the original torque bias, but electronic control took over automatically at certain times. It was super-sophisticated and super-costly stuff, but at a projected retail price of $125,000 per copy, Porsche could afford to be tricky.

Improvements over the original 911 platform abounded. The devilish suspension setup of the older car—which made the 911 infamously easy to spin—was abandoned in favor of racing-style double wishbones all around. Huge 956 brakes with ABS dealt with hauling the car down from what promised to be a considerable top speed. The entire underside was swathed in an aerodynamic, downforce-producing Kevlar belly pan.

The Gruppe B's engine was a tale unto itself. Based in murky pasts on the single-overhead-cam air-cooled, 911 powerplant, its evolutionary path split with the street car's in 1978. That was the year Porsche introduced water-cooled, four-valve, double-overhead-cam heads to the air-cooled cylinder barrels of the traditional 911—a setup originally used for the remarkable 935/78 Le Mans

126

car. Later developed into a 2.65 liter Indy car powerplant for an aborted Indy 500 effort, the engine found duty instead in the 956/962 prototypes that went on to dominate sports car racing for most of the eighties. This racing engine, in detuned form, went into the new all-wheel-drive prototype where it could conceivably find its way to the street.

The Gruppe B designation was soon dropped in favor of 959—the company's internal project number—as Porsche became more forthright with details of the car. For 1984, the company even entered a team of all-wheel-drive 911s in the Paris-Dakar Rally. The all-wheel-drive 911s were seen, and rightly so, as the first step in the development of raceworthy 959s. Aided by reliable Porsche 911 power and a typically ruthless level of preparation, the car driven by Rene Metge won the event outright.

Porsche's preparation for the race had been remarkable. Of the three cars it entered, the first two were racers and the third carried a team of "flying mechanics" to the race cars en route. Sensitive pieces like transmission parts were flown in nightly inside sealed plastic bags, ensuring that the Porsche parade could continue on the next day's leg of the race.

The next year's Paris-Dakar was a failure for Porsche, despite its running "prototype" 959s with full 959 bodywork, variable all-wheel-drive and a six-speed gearbox. To make up for that defeat, the first true racing 959s, called 961s, ran and won the 4,000 mile Pharoahs Rally later that year. Detuned

to take the lousy fuel available for the event, the 961 engines made "just" 370 hp. Set up for true racing-grade gasoline, that output was expected to top 700 bhp with ease.

The 959/961's ultimate performance came in the 1986 Paris-Dakar, a revenge event for Porsche. The result was a one-two finish, with the mechanics' car coming home sixth in the hands of Roland Kussmaul. The same year, Porsche entered a 961 road racer in the IMSA Daytona finale. Since the 961 didn't fit into any established IMSA classes, it ran as a demonstration car in the unheard-of GTX category.

By then the 959 street car was ready to roll. Its introduction in road-going form was delayed several times, in part because Porsche decided it should meet all existing emissions requirements—even those in California—with no horsepower loss.

By the time it was ready, though, its specifications were as impressive as Porsche could make them: 450 hp, electronically controlled infinitely variable all-wheel-drive, stressed Kevlar body panels, the lowest drag coefficient of any Porsche street product, speed-related electronic ride control and much, much more.

One interesting aspect of the 959 was its computer-controlled all-wheel-drive. Early on, Porsche replaced the viscous differential of the Gruppe B in favor of an in-house design using two ten-disc hydraulic clutches between the front and rear differentials to vary the torque split as needed. (A similar idea was seen in Opel's Xtrac transmis-

The 961 road racer didn't fit into any existing IMSA classes, so Porsche entered it in a one-car GTX cate-gory. There was no prize for running except experience and great publicity. *Road & Track*

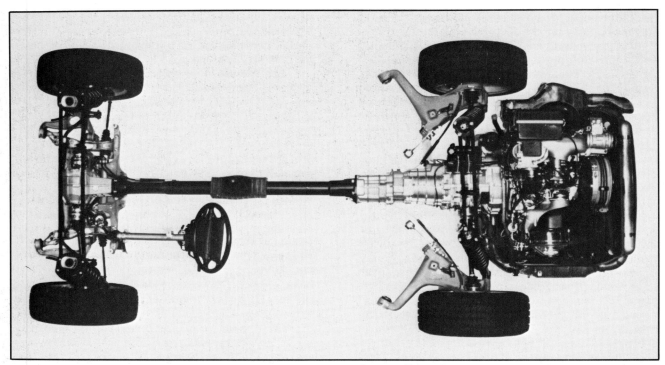

The Porsche 959's multiplate center and rear differentials found a less exclusive home in the mass-production Carrera 4. The Carrera 4's suspension, however, wound up closer to the 911's. *Porsche of North America*

sion system, but Porsche's was by far the more ambitious and sophisticated of the two.) In full rear-wheel-drive, the car would merrily kick its tail out through slow corners. As speed increased, so did the torque to the front axle, until it reached fifty-fifty for maximum straight-line stability.

Porsche's announced 200 plus production run for the 959 sold out before the first road car was assembled, despite the hefty price tag and long delay. Within a year some examples were already trading well beyond the million-dollar mark.

Jaguar XJ-220

The Birmingham Motor Show in 1988 had one indisputable star, a super-coupe Jaguar called XJ-220. The best thing about the Jaguar's introduction that day was that it completely stole the thunder from the show's expected star, Ferrari's twin-turbo F40. The two cars faced each other across an aisle, and the Jaguar stared the F40 down.

Rumors of such a machine from Jaguar had been floating around since long before Porsche's 959 existed. But the Birmingham show marked the company's official unveiling of what could still become the world's fastest production automobile.

The very appearance of the XJ-220 prototype is a small wonder. The program began when a select group of Jaguar employees, noticing the tremendous press the Porsche 959 and Ferrari GTO gained for their respective makers, embarked on an after-hours project to create a Jaguar super-car in a similar vein. Their goal was to produce a concept car capable of winning FISA Group B competitions against the Porsche 959 and the imminent evolution model of the Ferrari GTO.

Without the deep pockets of Porsche or Fiat (Ferrari's parent company) behind it, the super-Jaguar was at first strictly a labor of love. As it developed, however, Jaguar employees from top man Sir John Egan on down fell in line behind the project, and funds began to trickle in. If enthusiasm and moral support were enough to build a car, the XJ-220 would have been prowling the world's roadways in record time. As it was, FISA Group B racing was stillborn before production could be started and the car's future remained in doubt as the eighties came to a close.

In finished specifications, however, the XJ-220 looks too good to be true: mid-mounted V-12 engine, bonded aluminum magnesium monocoque construction, hand-hammered aluminum body panels, Ferguson Formula engineered all-wheel-drive and a theoretical top speed of 220 mph.

Best of all, the 220 is producible, whether Jaguar takes up the challenge or not. Its advanced construction follows simple race car procedures that would allow numerous, if costly, replicas. The engine, for example, is essentially a hopped-up version of Jaguar's street-going aluminum V-12, fitted with Zytec electronic engine management, four-valve double-overhead-cam heads and port fuel injection. A distributorless electronic ignition system makes use of twelve individual coils, one at each spark plug. No forced induction system would be needed to make its 500+ bhp, as none was used on the Group C and IMSA GTP winners that ran similar powerplants in competition.

For the driveline, a FF Developments viscous center differential is mounted between the longitudinally placed engine and gearbox. The all-wheel-drive layout feeds sixty-nine percent of the torque to a viscous rear differential directly beneath the gearbox and thirty-one percent through spur gears to a driveshaft nestled in the vee of the engine—the location formerly held by the auxiliary shaft off which production Jaguar V-12s drive their distributors. That shaft goes forward to a front differential that may feature viscous limited slip as well. With all that power and traction, 0–60 times are expected to be 3.5 seconds or less, definitely world-record territory.

The XJ-220 bears more than a passing resemblance to a GTP car, in no small part because the longitudinal V-12 pushes the passenger compartment well to the front of the already long chassis. This is no small machine: 202 inches overall. The wheelbase alone stretches 112 inches, and the width is a Testarossa like 78.7 inches at its widest point. Its width is part of the reason the car has a rather unimpressive drag coefficient of .38.

The main reason for the 220's high drag figure is its use of radical ground-effects tunnels to produce downforce. While ground effects don't add as much drag as spoilers do, any system that produces downforce equal to the force of gravity at top speed is bound to take its toll.

What could keep such a machine off the road? Simply money. In 1989, Jaguar was suffering from an acute shortage of funds, in part because of the poor pound-per-dollar relationship and in part because the company had to press ahead with the less-ambitious-though-still-exciting F-Type sports car program. The relatively high production F-Type is seen as a cornerstone of Jaguar's future line-up.

Jaguar doesn't have the resources to pour into the XJ-220 if it's to be used strictly for publicity. It will have to make money in its own right, which is almost unheard-of with a production run of 200 vehicles or so. To that end, every effort was made to keep initial costs down while keeping technology up. Subcontractors played an unprecedented role in the car's development, and much of their time

The Chevrolet Corvette Indy made its debut with a mid-mounted, longitudinal 2.65 liter Chevrolet Indy racing engine. With the racing engine, production possibilities seemed slim. *Chevrolet*

The long, low and lean Indy show car shares some of its design philosophy with the Oldsmobile Aerotech and Pontiac Banshee. The raised fenderline and roof section of the second-generation car did nothing to detract from its appearance. *Chevrolet*

was also donated after-hours by enthusiastic employees.

Rumor had it that the 959 delayed Porsche's regular production engineering by nearly two years. That will not be allowed with the XJ–220, even though in the long run the 959 probably paid off and then some in favorable publicity alone.

If Jaguar's beast is to see production, it will be built not by Jaguar but JaguarSport, the ambitious company owned equally by Jaguar and Tom Walkinshaw, operator of the highly successful Group C and IMSA GTP prototype efforts.

To make the car pay for itself, Jaguar figures a base price of $250,000 or more will be necessary. As more and more enthusiasts show interest, however, that price is raised almost hourly and could go as high as $500,000. While that seems a lot to pay for a new car, Jaguar knows that Porsche and Ferrari made a mistake by selling their own Group B specials at a loss, only to see them turn over to speculators for three and four times their purchase prices.

By late 1989, the go-no-go decision was still unmade, at least officially. Jaguar had even refused a number of $100,000 deposits offered by prospective buyers, a bad sign if production was to begin any time soon. But while such a project might be daunting to a company without Jaguar's pluck—Aston Martin's mid-engined Bulldog comes to mind—Jaguar has proved it can do what it sets out to do.

For the company that took just ten short years to go from death's door and dated goods to black

ink, customer satisfaction and a return to the winner's circle at Le Mans, anything's possible.

Chevrolet Corvette Indy

When General Motors acquired a controlling interest in the Lotus organization, its main goal was to add Lotus' advanced engineering staff to its own. Lotus' strong suit has always been forward-thinking design and engineering, and the General needed a dose of just that.

One of the first projects assigned to the English group was the design and construction of the next Corvette show car. This was no small task; from the start of the Corvette program, Corvette show cars have been GM's most exciting and advanced dream vehicles. Lotus had to design a show car that was up to being judged against everything from the first Corvette prototype (so successful it went directly into production in 1953) to the sleek, mid-engined Aerovette of the seventies.

The decision was made to take a technology-intensive approach to the new show car, starting with all-wheel-drive, four-wheel-steering, heads-up instruments and active suspension. Not only would this technology be included in the car, it would be developed into a drivable, user-friendly, fast package.

The decision was also made to transversely mount Chevrolet's Indy V–8, the quad-cam racing engine that went on to dominate Indy car competition. It was tantamount to taking a Formula One engine and tuning it for the street.

130

Chevrolet and Lotus didn't project a top-speed figure, only guesses. However, the production Corvette would see 155 mph with 245 hp; the Indy's engine was rated at 600 bhp, and the sky was the limit if Chevrolet decided to up that figure.

The Corvette Indy was first introduced to the public in January 1986 and was an immediate smash. Because it was based on a superexpensive racing engine, the crowds knew there was little chance of anything like the Indy going into production. But that changed when Chevrolet replaced the Indy engine with an LT5, the Lotus-designed, quad-cam, aluminum V–8 that later went into production with the ZR1 Corvette. By shoehorning the LT5 engine into the Corvette Indy's engine bay, Chevrolet was giving the signal that a limited production run of Corvette Indy replicas was a possibility.

Auto magazines picked up the gauntlet and challenged Chevrolet to do just that. Chevrolet responded with a second-generation show car geared more toward drivability, with higher wheel arches for better suspension travel and a raised greenhouse for more headroom.

Toward the end of the eighties, however, much of the optimism about a production version of the Indy began to fade. Chevrolet had been quiet about the project for too long, and it seemed the company was willing to simply let this Corvette show car drift into oblivion as the earlier Aerovette and Astro II had. Enthusiasts who had dreamed of

Computer goodies abounded in the Corvette Indy's interior. Rearview video camera and passenger's computer terminal are still probably the stuff of dreams, but the heads-up display has already arrived in other GM vehicles. *Chevrolet*

seeing the Indy on the road remembered that this was GM they were talking about, not Ferrari. The running joke was that even the ZR1 had only been built because one day Chevrolet's performance people held a planning session and gave the company's other executives the wrong room number.

Embodied in the Indy, at least, are a number of technological advances that the next-generation Corvette might well employ. Its four-wheel-steering,

The Pontiac Turbo Grand Prix marked GM's first production heads-up display. Pertinent information is projected onto the windshield so the driver needn't take his eyes off the road. *Pontiac*

drive-by-wire electronic throttle system, heads-up instruments and active suspension all point to production parts of the future. The Indy's all-wheel-drive, however, would be difficult if not impossible to adapt to the Corvette in its current front-engined form. There's so little room in the production car's crowded engine bay that finding space for driveshafts and a front differential would be a major task.

But the almost insane size of the ZR1's rear tires—12.4 inches wide—demonstrates that traction for acceleration is becoming a real concern with the Corvette. If Chevrolet intends to keep producing super-high-output versions of America's only sports car—and it does—some form of all-wheel-drive system could become almost man-

Perhaps the next-generation Corvette will use a mid-engine? Or a shorter, force-fed V-6? Or a compact front differential with shaft-through-sump front drive axles? Chevrolet has never mentioned all-wheel-drive in connection with a production Corvette, but enthusiasts are staying tuned. In rear-wheel-drive form, the Corvette's already one of the world's best sports car. All-wheel-drive would only make it better.

Nissan MID4

In the name Nissan MID4, MID stands for mid-engined, and 4 means four of almost everything else: four-wheel-drive, four-wheel-steering, four camshafts, four valves per cylinder, four-wheel vented-disc antilock brakes, four coil-over shocks on the front suspension. OK, so there were only three viscous differentials, two seats and two intercooled turbochargers, and one torque tube to the front final drive, but who's counting?

The first-generation MID4 raised a ruckus in the reawakening performance world of the mid-eighties, but despite its impressive technical specifications, its uninspired styling left many people cold. Nissan simply whipped up a second-generation car to answer their complaints, throwing in a good deal of added drivability and interior room to heighten the appeal.

The Nissan MID4 will probably never see production; Acura's visually similar NS-X will be the first and probably only Japanese supercar through at least the early nineties. But much of the MID4's technology has already hit the showroom. The V-6 in the current 300ZX, for example, is actually a second-generation development of the MID4 block.

Alongside the MID4, Nissan introduced its sleek CUE-X four-door. This clean (.24 Cd) and pleasing sedan pulled out all the high-tech stops, making the MID4 look positively backward. Torque-variable all-wheel-drive, speed-sensitive four-wheel-steering, electronic spoilers that deployed at speed or under braking or during rain, laser-radar

The rear styling of the Nissan MID4 reflected the mass of machinery mounted in the car's tail. Vents for cool-ing and a tall rear deck line were reminiscent of the Ferrari 308/328. *Nissan*

The first generation of the Nissan MID4 show car combined elements of Lotus, BMW and Ferrari styling with distinctly Japanese touches. The overall effect was a qualified success. *Nissan*

Although the MID4 was intended for show duty, Nissan engineers paid great attention to its aerodynamics. A wind tunnel test sends visible smoke jets over upper surfaces to check wind resistance. *Nissan*

The clean, understated lines of the CUE-X garnered as much attention as did the high-tech goodies inside.

The car's relatively unadorned shape reflected Nissan's increasingly worldly outlook. *Nissan*

crash sensors that activated the antilock brakes, variable-nozzle turbocharging and a host of other goodies made the CUE-X a hardware lover's dream.

Despite all the high-tech nuances, the CUE-X body shape was perhaps Japan's best. Highly reminiscent of the Ferrari Pinin four-door show car, the CUE-X presented an understated but completely elegant face to the world. Except for a single, off-center hole in the car's nose, it remains the most mature and well-balanced design yet from Japan.

Peugeot OXIA

Peugeot, a class act in high-performance all-wheel-drive with its 205 and 405 based rally cars, decided to join the all-wheel-drive show car ranks with the OXIA, a mid-engined street-legal racer with a top speed far in excess of 200 mph.

The OXIA's intercooled, twin-turbocharged V-6 produced 670 hp despite its supercompact packaging, and all-wheel-drive was necessary to get its considerable power to the road. From the engine, a twenty-five percent front, seventy-five

At home on a banked racetrack, the Peugeot OXIA looked like a cross between a Le Mans entry and a 405

street car. Slap on a new nose and some sponsor decals and you'd have a nice FISA racer. *Peugeot of America*

The dashboard of the OXIA held solar cells wired to an interior ventilating fan. When the car was not cruising at 200+ mph, it could get hot under that big windshield. *Peugeot of America*

percent rear torque split tamed driving forces with the help of four-wheel-steering, ABS and computer control throughout. Like the Porsche 959, the Peugeot OXIA also featured an onboard monitoring system for tire pressure, warning the driver in advance if there was going to be tire-related trouble on a high-speed run.

The carbon fiber body, computer-controlled aerodynamic aids and computer-intensive passenger compartment of the OXIA may still be a long way off in production models. But its advanced driveline components are another matter. Should Peugeot decide to produce a street-bound supercar with anywhere near the horsepower of the OXIA, a similar all-wheel-drive system is sure to come along for the ride.

MG EX-E

One appealing show car of the late-eighties bumper crop came from the Austin-Rover group of England, the same people responsible for the MG Metro 6R4 rally car. However, while the 6R4 was generally considered an ugly duckling, the EX-E was quite the opposite.

Clean, simple and sleek, the EX-E was one of the best looking shapes to come out of England since the E-Type Jaguar.

Featuring a 410 hp version of the 6R4's normally aspirated V-6, the EX-E leapt to 60 mph in less than five seconds. Its charm, however, was in its clean overall shape—good for an astonishing .24 Cd—and the relatively straightforward design of its underpinnings. The diminutive dream car's all-wheel-drive drivetrain has already been proven in the 200 odd 6R4s driving Europe's roadways, so getting the machine into production would be relatively easy. And with its straightforward, nontur-

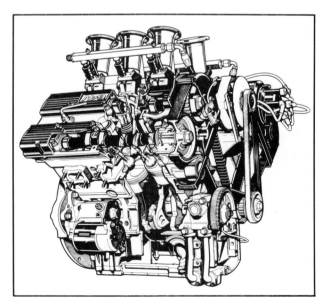

The MG EX-E/6R4 engine traced its lineage to an aluminum-block Buick V-8 later built by Austin-Rover. Two cylinders were chopped off, four cams and electronic injection appeared for rally duty. *Kermish-Geylin Public Relations*

The jet fighter styling of the EX-E yielded a low .24 coefficient of drag. If Austin-Rover wants to get into the exotic car business, this is the way to do it. *Kermish-Geylin Public Relations*

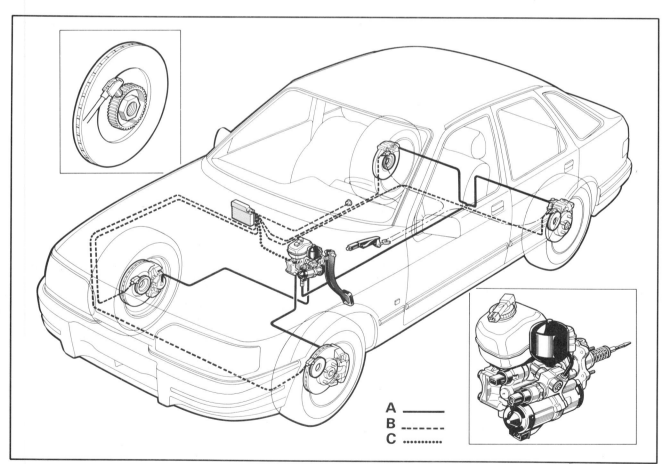

A ———
B - - - - -
C ·········

Hydraulic lines (solid lines) release brakes when an electronic signal from the control unit (dotted line) tells the master cylinder to reduce pressure. Inputs from wheel sensors (dashed lines) talk to control computer. *Ford of Europe*

bocharged engine, keeping it on the the road would be as simple as—well, at least simple for an English mechanic.

At the end of the eighties there was still no word from Austin-Rover that such a plan existed. Austin-Rover, like Jaguar, doesn't have R&D money just lying around in bags. But the overall appeal of the EX-E and the tremendous interest shown by auto enthusiasts could well change its mind, if not toward the EX-E itself then at least toward something much like it.

MG's heritage has always rested with fast, simple and good-looking sports cars at a reasonable price. A machine like the EX-E would almost cer-

tainly be priced in the lower echelons of exotic territory, but on a bang-for-the-buck scale it could be a world beater.

Ferrari 408

It might well have turned out to be a production prototype, but a staff shakeup has most likely relegated the 408 to permanent dream car status, taking its all-wheel-drive with it. The innovative 408 placed a four-liter Ferrari V-8 longitudinally behind the driver and offset slightly to the passenger's side. A forward driveshaft was fed through a double hydraulic coupling similar in principle to an automatic transmission. Torque was split 29.3

Ferrari's 408 concept/show car put plenty of the company's advanced thinking into a single vehicle. It was the first proof of any all-wheel-drive aspirations in Maranello. *Alcan Aluminum Ltd.*

In addition to all-wheel-drive, the 408 featured speed-sensitive ride height and a 300-bhp aluminum V-8. The second-generation car added Alcan's Aluminum Struc-tured Vehicle Technology—a bonded-aluminum central cage—to the package. *Alcan Aluminum Ltd.*

At low speeds, four-wheel-steering on the two-wheel-drive Honda Prelude turns front and rear wheels in opposite directions. Wheels turn in the same direction when the car is traveling faster. *Honda North America*

percent front, 70.7 percent rear, reflecting the tail-heavy layout of the automobile.

The 408 also boasted a bonded aluminum central structure, foam-filled plastic body skins and hydraulically actuated semi-active suspension. Whether any of its innovative features will appear soon in Ferrari's road cars is anyone's guess. The 308/328 GTB replacement has already arrived in two-wheel-drive form, much to the surprise of some pundits; many predicted the new car would be patterned after the 408.

Advanced technologies in concert with all-wheel-drive

Manufacturers are generally quiet about their plans for further improvements in all-wheel-drive, preferring to spring their advances on the public right before production. This may be because their marketing divisions have told them that all-wheel-drive continues to confuse a large section of the buying public. Dealers have enough trouble explaining the virtues and operation of systems that already exist; getting mired down by those that may be coming would make the situation even worse.

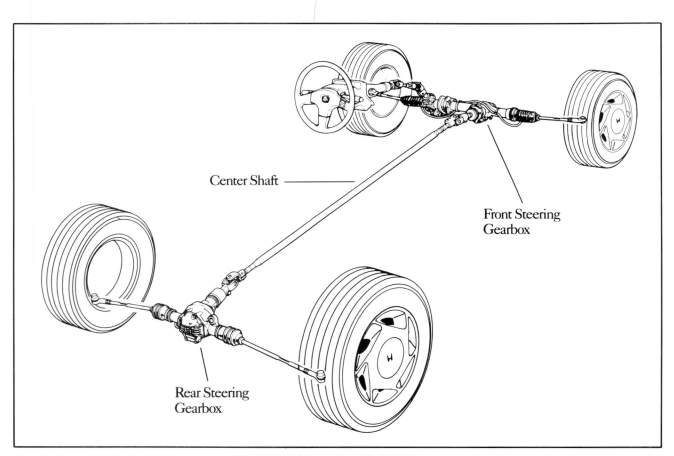

Honda four-wheel-steering uses a separate steering gearbox for the rear axle, driven off a long U-jointed shaft. Future systems may use electric motors at the rear. *Honda North America*

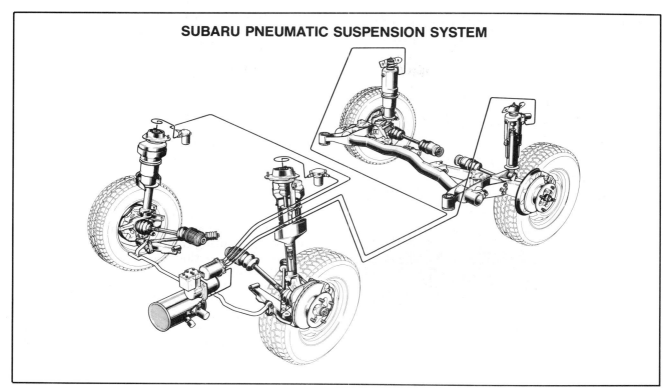

SUBARU PNEUMATIC SUSPENSION SYSTEM

The pneumatic Subaru suspension system uses bladderlike springs at each wheel. Individual wheel sensors provide constant load leveling while a dashboard switch raises overall ground clearance on demand. *Subaru of America*

Still, much of the future of all-wheel-drive hardware is obvious. Higher degrees of control are in the pipeline, and applications will continue to spread. In the near future, all-wheel-drive may be at least an option on more automotive platforms than not.

But to get a good grip on tomorrow's all-wheel-drive offerings, it perhaps pays more to look at technologies that can be employed with all-wheel-drive systems to further enhance their performance.

Capitalizing on the improved stability of all-wheel-drive, for example, many manufacturers now include antilock braking systems on all their all-wheel-drive offerings. ABS employs sensors to detect wheel lockup, and momentarily releases braking force to ensure that a wheel doesn't skid. When all four wheels are solidly locked together—as in an all-wheel-drive with locked center and rear differentials—ABS is generally deactivated to avoid adding undue stress to the driveline. (Since the four wheels are effectively tied together anyway, ABS is not so important.) With systems that allow limited differential action at all times—such as Torsen differentials and viscous couplings—however, ABS remains active constantly.

Electronic traction control is another feature that will probably appear soon in all-wheel-drive offerings. A few two-wheel-drive cars (Volvo, Mercedes, Cadillac, and others) already offer it. Traction control is basically the reverse of ABS. Again, sensors detect any large speed difference between wheels, but in traction control a brake is applied or engine power is momentarily throttled back when one wheel begins turning *faster* than the rest, allowing that tire to get a bite on the pavement.

Once the sensors for ABS are in place, installing traction control becomes a relatively simple matter in two-wheel-drive vehicles, since the speed of the drive wheels can easily be measured against the speed of the driven wheels. With all-wheel-drive, however, a decision must be made as to whether a traction problem actually exists or if the car is simply accelerating.

This is not an insurmountable problem: The electronic control unit that oversees traction control could be programmed to accept only a certain increase in wheel speeds over time, or it could take input from a potentiometer and correlate it to the wheel sensor's information. Once this trouble is overcome, the only real difficulty with traction control is designing efficient and safe throttle linkages. That won't be a consideration when drive-by-wire throttle systems become common.

Also adding to the inherent stability of all-wheel-drive will no doubt be four-wheel-steering—

The compact Borg-Warner modular all-wheel-drive system uses a chain to take power off to the front of the car. The system's computer controls integrate with existing sensors to make installation easier. *Borg-Warner*

At the end of the eighties, many manufacturers had already made steps toward supplying suspensions with variable ride characteristics—suspensions that firmed up and relaxed according to road conditions, driver input, speed and even sonar road sensors. This sort of semi-active technology is an extension of the ride-adjustable suspensions of the sixties and earlier—for example, the externally adjustable Spax shock or the famous hydraulic suspensions from Citroen. Subaru's optional pneumatic dampers work on the same principle, allowing a number of ride characteristics to come out of one set of hardware. Such semiactive or reactive suspension systems are quite limited in their abilities, however.

Fully active suspensions are another matter entirely. Instead of using conventional springs, torsion bars and so on to suspend the car above the road, they carry the vehicle on four hydraulic cylinders under computer control. The cylinders are infinitely and immediately adjustable. Over large potholes, they instantly mimic conventional soft suspension settings. In turns or high-speed maneuvers, they firm up race car hard.

Active suspension is expected to make its debut in the early to mid-nineties, the apparent leader in its development being Lotus and, by default, General Motors. A Cadillac, Lotus or Corvette is the most likely recipient of the world's first true active suspension system, and active suspension with all-wheel-drive won't be far behind.

Future all-wheel-drive hardware

All-wheel-drive hardware is in for improvements in the years to come, and some of those are well known. Borg-Warner, the company that developed Jeep's pioneering Quadra-Trac viscous coupling driveline, has already introduced an electronically adjustable modular all-wheel-drive system that can easily be tailored to accept a variety of electronic controls. Completely overseen by a programmable computer, the variable-torque-split Borg-Warner system will work in concert with the existing sensors and computers of any automobile to deliver whatever characteristics the designers desire. This system makes adding all-wheel-drive to an existing platform easier (and therefore less expensive) than ever.

Torque distribution for the Borg-Warner box is handled by means of an electromagnetic clutch pack, so allowing either the computer or driver to select torque split is a simple matter. The compact Borg-Warner system can also be easily adapted to Borg-Warner's or other manufacturers' transmissions, making installation easier still.

Jaguar patented a unique system in 1987 that promises to appear on one of its own street vehicles in the future. This mechanical all-wheel-drive biasing system varies torque split from gear to gear, starting with a 1:3 rear bias in first gear and ending

perhaps better described as *active* four-wheel-steering, since simple inertial systems to allow toe changes and so forth in cornering are not really new or exciting. Already an almost mandatory addition on show cars, four-wheel-steering was recently introduced in limited production by Honda and Mazda.

As the name suggests, the idea behind four-wheel-steering is to steer the rear wheels in addition to the front wheels. At low speeds, the direction of rear steer is opposite that of front steer, aiding in low-speed maneuvers such as parking. At high speeds, all the wheels turn in the same direction for added stability and faster response. Fast lane changes in a four-wheel-steer vehicle are stunning; the car simply leaps ten feet to the left or right, without the oversteer or understeer usually inherent in such a maneuver. Expect many of the next generation of supercars to employ four-wheel-steering in addition to all-wheel-drive, and a growing number of less exotic offerings to feature it as well.

Finally, the most exciting technology to appear in concert with all-wheel-drive may be active suspension. Any conventional suspension must strike a compromise between stiffness for quick handling and softness for a smooth ride. Active suspensions constantly readjust themselves for given conditions to produce both at the same time.

with about a 3:2 split in fifth (assuming a five-speed box is used). The theory is that the rear bias will aid acceleration during takeoff and in throttle steering through tight corners, while the front torque bias adds stability at high speed.

The Jaguar system is quite simple and may cost less to build and sell than electronic torque-varying devices. Whether it will be advanced enough to suit Jaguar's forward-thinking profile is another matter, however. In that case, the licensing fees alone could pay for R&D on more sophisticated drive systems for Jaguar's own cars.

Another all-wheel-drive system that will no doubt appear shortly is the viscous torque splitter. Similar to the viscous transmission, the torque splitter goes one step further in eliminating traditional mechanical components. Where the viscous transmission's output goes to a regular differential, the torque splitter combines the functions of both components in a single housing. Torque is fed to what's essentially one large viscous coupling rotating in the same plane and direction as the wheels. Two separate shafts feed power out from either side, each one coupled to its own set of discs operating in the viscous coupling housing.

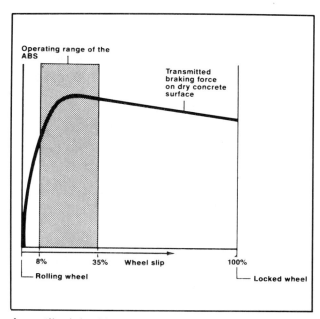

An antilock braking system releases braking force as traction falls off with wheel slip. By keeping the brakes at their most efficient level, stopping distance decreases. *Toyota*

The Mitsubishi HSR (for High Speed Research) concept vehicle promised to lend lots of ideas to Mitsubishi's later sports cars. The technology inside filtered down to less exotic models as well. *Mitsubishi of America*

Completely reliable and straightforward, the torque splitter is going to take over a share of the all-wheel-drive market. Who'll be first to use it is anybody's guess.

Audi, while quiet on its own all-wheel-drive advancements for the near future, has released information on a program that will take the research and development of such systems a big step further. At an official cost of $18 million (though it has probably cost much more than that), Audi has built the world's first wind tunnel designed specifically for all-wheel-drive automobiles. The Neckarsulm facility uses a four-wheel dynamometer that allows all driven wheels to operate at speeds up to 187 mph. Since four driven wheels have a noticeable effect on airflow at speed, the new wind tunnel allows Audi to more accurately predict how a given platform will work in concert with all-wheel-drive. Having the tunnel won't hurt the firm's GTO racing program, either: the dynamometer is also capable of dealing with engine outputs up to 500 bhp.

Some even predict that the basic design of all-wheel-drive will change dramatically by the year 2000 or so, doing away entirely with conventional drive systems for the secondary axle and substituting electric or hydraulic motors that will be activated only when all-wheel-drive traction is needed. This prediction, while interesting, overlooks one crucial thing: Such a motor would have to be smaller, lighter and less expensive than existing and future conventional drivelines. With the advanced state of all-wheel-drive we have already, such a development seems a long way off at the least.

Most likely, we'll see a number of evolutionary changes in all-wheel-drive hardware throughout the nineties. Small steps forward will lower the cost, increase the controllability and refine the action of all-wheel-drive from year to year. The only sure bet is that Harry Ferguson's dream of the fifties—the common application of "invisible" all-wheel-drive for safety and performance—will continue to gain speed.

The Mitsubishi 3000 GT, pictured, and the similar Dodge Stealth were scheduled to appear as 1991 models. Their all-wheel-drive, electronically controlled suspension and engine goodies first appeared in the HSR. *Mitsubishi of America*

Index